Random Signals and Systems

Random Signals and Systems

RICHARD E. MORTENSEN
University of California at Los Angeles

A WILEY-INTERSCIENCE PUBLICATION
JOHN WILEY & SONS
NEW YORK CHICHESTER BRISBANE TORONTO SINGAPORE

Library of Congress Cataloging in Publication Data:

Mortensen, Richard E.
Random signals and systems.

"A Wiley-Interscience publication."
Bibliography: p.
Includes index.
1. Signal theory (Telecommunication) 2. Stochastic processes. 3. Random variables. I. Title.

TK5102.5M68 1987 621.38′043 86-19007
ISBN 0-471-84364-4

Printed in the United States of America

10 9 8 7 6 5 4 3 2 1

To Guruji and Swamiji

Preface

This book was written to serve as a textbook for either a senior-level introductory course in stochastic processes or a first-year graduate-level follow-up course. The prerequisites for the senior-level course are courses in mathematics usually required of a senior majoring in electrical engineering, namely, matrix algebra, differential equations, and Laplace and Fourier transforms, plus a one-quarter course in probability. Although no additional specific prerequisites are presumed for the graduate course, it is hoped that graduate students will possess a significant intangible asset sometimes called "mathematical maturity." The courses for which this book serves as a text are prerequisites at UCLA for graduate courses in control systems engineering and communications systems engineering. The book is intended to provide the basic knowledge needed for learning to design analog communication systems and linear control systems operating in an aleatory environment as well as for gaining an understanding of standard digital signal processing techniques.

These courses in stochastic processes in the Electrical Engineering Department at UCLA are the descendants of similar courses that have typically been taught for the past 30 years using as texts such books as Davenport and Root (1958) or Papoulis (1965), which are listed in the reference section. During that time, there have been two historical developments that have had contrary impacts upon the way the subject of stochastic processes now needs to be taught. The first of these is the increasingly high level of mathematical sophistication that one encounters in research papers in the *IEEE Transactions* and many other journals pertinent to the subject. The second is the development and proliferation of the microprocessor and associated software.

A consequence of the first development is that a Ph.D. student in electrical engineering who expects to do cutting-edge research and make a theoretical contribution in a dissertation will have to know measure theory and functional analysis. A consequence of the second development is that recent textbooks for undergraduates on digital filter design make very minimal assumptions concerning the reader's background in calculus, but do presume some familiarity with discrete mathematics. In courses such as the one for which this textbook is designed, we are confronted with the dilemma of taking a student whose undergraduate preparation largely reflects the second trend and attempting to prepare that student for a career in a graduate school curriculum oriented toward the first trend.

Just for the sake of illustration, a recent text that exemplifies the first trend is the excellent book by Wong and Hajek (1985). A contemporary text exemplifying the second trend is the enjoyable book by Williams (1986).

I believe that a student who proposes to do serious work in stochastic processes at the Ph.D. level must take a course in real analysis from the mathematics department, followed by a course in functional analysis using a text such as the one by Balakrishnan (1981). Therefore, I have not endeavored to introduce the student to measure theory in this book. I have attempted to set him or her thinking along appropriate lines, by introducing probability in terms of set theory, by calling the probability set function a "probability measure," and by dropping hints here and there. For the same reason, there is no mention, except here, of the Ito stochastic calculus and associated topics in this book. On the other hand, I have ventured to discuss Hilbert space openly and without embarrassment. Although this choice may appear largely idiosyncratic, it was based on my experience in terms of what seems to work and what does not. I do believe it is desirable to introduce the Karhunen–Loeve expansion in a course such as this one, because it is part of the theoretical basis for such successful practical contributions as the Viterbi decoding algorithm. While discussing Karhunen–Loeve, some information on Hilbert space theory is very useful. It also helps in providing an interpretation of the meaning of the innovations process in Kalman filtering theory.

In order to accommodate our computer-oriented undergraduates, Chapter 2 includes a review of the Gaussian distribution in one and two dimensions and an exploration of some of its properties. This incidentally gives me an opportunity to check out the student's ability to do matrix manipulations and to evaluate multiple integrals. Chapter 4 meets the students on their level by discussing finite length random sequences, and in Appendix 2 we provide a computer program that will generate such sequences for students to investigate as they like.

The essential core of the undergraduate course is the material in Chapters 2, 4, 5, and 6. Depending upon the time available and the refractoriness of the students, material from Chapters 3, 9, and 10 can be introduced. The graduate course reviews all of that material and then takes up the discussion of the Hilbert space of second-order random variables from Chapter 1 and the Hilbert space of square integrable functions from Chapter 8, and then proceeds into the presentation of the Karhunen–Loeve expansion. Again depending upon time and opportunity, we can cover the properties of the conditional multidimensional Gaussian density and the introduction to estimation theory from Chapter 3, the state-space theory of dynamic systems from Chapters 7 and 10, and the introduction to Kalman filtering theory in Chapter 11.

The book was deliberately put in the form of a smorgasbord of topics, for maximum flexibility. The style is informal and discursive in order to keep the attention of most students. The theorem–proof format is used only in a few places where it seems particularly desirable to summarize the development and provide a concise statement of results.

Having now tried at some length to explain what I was trying to do, what I think I did, and why I made those particular choices, it is appropriate to express my gratitude to some of my colleagues for facilitating my efforts. I wish to thank Professors A. V. Balakrishnan, Jack Carlyle, and Stephen Jacobsen, all former chairmen of the former Department of System Science, for creating an environment that stimulated the genesis of this book and for their incredible patience with my idiosyncratic behavior. Further thanks go to Professors C. R. Viswanathan and Fred Allen, the former and the current chairman, respectively, of the Department of Electrical Engineering, which largely absorbed the Department of System Science, for creating the nurturing climate that enabled this book to become a reality. Additional gratitude is due my colleagues on the faculties of those two departments, whom I will not mention individually for fear of overlooking someone, for many enjoyable discussions over the years. I am deeply indebted to my students over the past 20 years, who taught me valuable lessons about education.

Finally, my sincerest thanks go to Sophie Spurrier for typing the entire manuscript and enduring the process of making seemingly endless corrections and improvements.

RICHARD E. MORTENSEN

Los Angeles, California
November 1986

Contents

Chapter 1

Discussion of Probability and Stochastic Processes 1

Introduction 1
Probability 1
Random Variables 4
Independence and Conditional Probability 9
The Hilbert Space of Second-Order Random Variables 10
Random Processes 13
Problems 18

Chapter 2

The Gaussian Distribution in One and Two Dimensions 21

The One-Dimensional Gaussian Distribution 21
The Two-Dimensional Gaussian Distribution 23
Integration Over a Quadrant 25
Problems 29

Chapter 3

The Multidimensional Gaussian Distribution 31

The Joint Density Function 31
The Conditional Density Function 33
Matrix Inversion Lemma 34
Conditional Mean and Covariance 38
Significance of the Conditional Mean: Introduction to Bayesian Estimation 40
Problems 44

Chapter 4

Finite Random Sequences 49

Introduction 49
The Successive Viewpoint 49
The Simultaneous Viewpoint 55
Lower Triangular Matrices and Causality 58
Problems 58

Chapter 5

Stationary Random Sequences 61

Mean and Covariance Functions 61
Example: White Noise Input to Discrete-Time System 63
Power Spectral Density and Bochner's Theorem 68
Digression: Review of Constant Parameter Discrete-Time Deterministic Linear System Theory 71
Input–Output Relations for Spectral Densities 74
Factorization of Rational Spectral Densities 76
Problems 78

Chapter 6

Continuous-Time Stationary Gaussian and Second-Order Processes 81

Introduction 81
Covariance and Spectral Density Functions 84
Laplace Transforms and Linear System Theory 87
Input–Output Relations for Stochastic Processes 88
Spectral Factorization and the Paley–Wiener Criterion 90
Ergodic Processes 93
Power Spectra for Deterministic Signals 97
Problems 100

Chapter 7

Nonstationary Continuous-Time Processes 103

Introduction 103
State Space Models 104
Time-Varying Systems 107
Problems 111

Chapter 8

Additional Topics in the Study of Continuous-Time Processes 115

Introduction 115
Hilbert Space of Time Functions on an Interval 116
The Karhunen–Loeve Expansion 120
Example: Karhunen–Loeve Expansion of Brownian Motion 124
Other Orthogonal Expansions 126
Example: Narrow-Band Noise in a Communication System 131
The Uncertainty Principle 136
Problems 142

Chapter 9

Linear Systems in Conjunction with Memoryless Nonlinear Devices 145

Introduction 145
The Square-Law Detector 145
The Wiener–Volterra Series 154
Modulation, Complex Waveforms, and Analytic Signals 160
The Radar Uncertainty Principle 164
Problems 169

Chapter 10

Nonstationary Random Sequences 173

Scalar-Valued Sequences 173
Vector-Valued Sequences 177
Calculation of Mean Vector and Covariance Matrix 180
Time-Varying State Space Models 183
Problems 185

Chapter 11

Discrete-Time Kalman Filtering 189

Introduction 189
Problem Formulation 190
$\mathbf{LDL}^T$ Factorization, the Innovations Sequence, and the Update Formula 192
The Markov Model and the Propagate Formula 198
The Kalman Filter Equations 199
Using the Filter 204
Problems 206

Appendix 1

Triangular Factorization of Covariance Matrices 209

Appendix 2

Statistics of Monte Carlo Simulation 217

A Random Number Generator 217
Estimation of the Parameters of a Distribution 218
Statistics of Estimators 221
Numerical Results of a Computer Experiment 224

References 227
Index 229

Chapter 1

Discussion of Probability and Stochastic Processes

Introduction

The purpose of this book is to present some particular topics from the theory of stochastic processes which have found applications in control and communications engineering. The book has been written on the assumption that the reader has already had an introductory course in probability theory. Nevertheless, for a variety of reasons it seems appropriate and useful to begin with a review of that subject.

In this chapter we provide a review of the main ideas from probability theory that will be needed in understanding the material in this book. Beyond that, we will introduce one or two ideas which will probably be new to the reader, such as the Hilbert space of second-order random variables, that also will be handy to have available. Finally, after we define some terms and develop some concepts, we will explain what we hope the student will acquire from studying the material in this book, and provide a brief survey of the task to be undertaken.

In order to do that, we will provide a tentative definition of the term "stochastic process," as well as a brief discussion of certain kinds of stochastic processes which will be encountered again subsequently. The end of the chapter also contains a short statement explaining why the book has been written the way it has.

Probability

It is widely agreed that a good way to study probability theory is to base it on set theory. We will approach the subject from that standpoint. The term "set" is, in very rigorous treatments, considered to be an undefined concept which includes certain properties that are assumed in the initial axioms upon which the whole subject is based. Intuitively, a set is a collection of objects. In probability theory, these "objects" are *elementary events*. In set theory, the set of all the objects with which one intends to deal is taken as the *universal set*. In probability theory, the universal set is called the *sample space*.

Suppose one does an experiment in which the element of randomness is known to play a role. For example, conduct a survey by selecting some category of people and asking them questions, or make repeated measurements of some physical variable under circumstances where experimental error is known not to be negligible. Such an experiment is sometimes called a *random experiment*. It is not the structure of the experiment that is random; instead, randomness refers to the fact that the outcome cannot be predicted precisely in advance.

The *statistics* of the experiment refers, at the most primitive level, simply to the data itself. On a more refined level, "statistics" also refers to certain properties the data is found to have after subjecting it to some numerical processing. Probability theory is used to analyze such a random experiment. It is used to decide what kind of numerical processing is appropriate for the data and what kind of statements one can make with confidence concerning the statistics. Even more basically, probability theory is used to determine how the experiment should be structured so that one *can* make meaningful statements with confidence.

In performing such an analysis using probability theory, it turns out to be a disadvantage to have a sample space that is too large or too small. Therefore, the choice of sample space is usually tailored to the experiment in question. For example, suppose the experiment is to flip a coin 10 times, and record the outcome of each flip, that is, whether it is heads or tails. A sample space with only two points in it, heads and tails, is too small and is actually not useful. A sample space with infinitely many points in it is certainly large enough. The problem is, it is so large as to be unwieldy, and it may lead one into mathematical distress of a kind that one prefers to avoid if there be a way of avoiding it.

The sample space for the above experiment which turns out to be "just right" is the set of all *binary sequences of length 10*. There are $2^{10} = 1024$ of these, so this sample space contains 1024 points. Each point is an "elementary event," that is, a complete sequence of 10 flips. A single flip is *not* an elementary event.

In doing mathematical probability theory this way, a numerical probability would *first* be assigned to *each elementary event* (each sequence of length 10). The value of the probability assigned to each event must be a real number between 0 and 1, and the sum of the values over all 1024 points of the sample space must be exactly 1.

At this juncture, we can look at various subsets of the sample space, for example, the subset consisting of all sequences having heads occur on the first flip. The sum of the values of probability over all of the points in this subset is, *by definition*, the probability of getting a head on the first flip. If that number agrees with what you intuitively feel ought to be the case, then you may say that your coin-flipping model is *realistic*. On the other hand, if that is not the value that you think the event of getting a head on the first flip should have, then you must change the probabilities assigned to the elementary events until things come out the way you want them to.

Probability theory will show you how to make calculations from your mathematical model concerning the probabilities of various events. It is up to *you* to take the responsibility for deciding whether or not the model is realistic. If you test it in situations where the correct answer is already known, and the model gives you the correct answer there, then you may feel confident in trusting it in situations where the answer is unknown.

Let us now give some precise mathematical definitions. The fundamental entity that we require in order to use probability theory is a *probability trio* $(\Omega, \mathscr{A}, P)$. The first member of the trio, Ω, is the *sample space*, which may be either finite, countably infinite, or uncountably infinite. The second member of the trio, $\mathscr{A}$, is the *algebra of admissible subsets of* Ω, also called the *algebra of events*. The third member of the trio, P, is the *probability measure* defined on $\mathscr{A}$. That is, P is a set function. Its argument is one of the sets that belongs to $\mathscr{A}$, and its value is a real number between 0 and 1.

If Ω is a finite set, then $\mathscr{A}$ is simply the collection of *all* subsets of Ω, the so-called *power set* 2^{Ω}. If Ω is an infinite set, it is not possible in general to assign a probability to every one of its subsets in a consistent way without encountering mathematical difficulties. Therefore, the family of subsets of Ω to which probabilities are assigned has to be specified. That is what $\mathscr{A}$ is. Its members obey the rules of Boolean algebra with respect to the operations of union, intersection, and complement.

With these agreements in force, the only conditions that the set function P must satisfy in order to be a probability measure are the following:

1. $P(\varnothing) = 0$ where $\varnothing$ = empty set
2. $P(\Omega) = 1$
3. $P(A) \geq 0$ for every A in $\mathscr{A}$
4. If $A_1, A_2, \ldots$ are disjoint members of $\mathscr{A}$, then

$$P\left(\bigcup_{k=1}^{\infty} A_k\right) = \sum_{k=1}^{\infty} P(A_k)$$

Random Variables

In addition to the function P defined on $\mathscr{A}$, we also consider functions defined on Ω itself. Any such function is called a *random variable*. If the value of the function is a real number, it is called a real random variable; if the value is a complex number it is called a complex-valued random variable; if the value is a vector in R^n, it is called a vector-valued random variable; and so on. It is customary to abbreviate "random variable" by r.v.

If the set Ω is infinite, then in order to avoid mathematical distress we have to ban certain pathological functions. It is very unlikely such a function would arise in most applications, but we will include this restriction for the sake of precision. Let us explain it further.

The class of admissible random variables must agree with our algebra of admissible sets. We will explain what "agree" means for real r.v.'s; the extension to more general r.v.'s is a technicality. If $X(\omega)$ is a real r.v., then we want to discuss the probability that the value of X falls in some interval I of the real line. In order to do that, we have to be dealing with an event. Therefore, define

$$X^{-1}(I) = \{\omega \in \Omega\colon X(\omega) \in I\} \tag{1}$$

The symbol $\in$ means "belongs to." It suffices for this condition to consider only the class of semi-infinite intervals of the form $I = (-\infty, a]$, for every real number a. If for each a, the set $X^{-1}(I)$ is a member of $\mathscr{A}$, then X is an admissible r.v.

Under those circumstances, we are assured that the probability $P\{-\infty < X \leq a\}$ of the event that X is less than or equal to a is well defined. We give this probability a special name. Since it is a function of the parameter a, we call it the *distribution function for the r.v. X*. It is denoted by $F_X(a)$. In symbols:

$$F_X(a) = P\{-\infty < X \leq a\} \tag{2}$$

Under appropriate circumstances, the distribution function $F_X(a)$ turns out to be differentiable with respect to the parameter a. This will happen only when the sample space Ω is uncountably infinite. In those cases it is convenient to work with the *probability density function*, defined as the derivative of F_X. It has become a common practice to use the same letter for the argument of this density function as is used to designate the random variable itself. Although this system may be used without confusion by those proficient in the subject, for students trying to master the fundamentals it is misleading and confusing. In this book we will always use a capital letter for random variables. The parameter in the density function will then be the corresponding lowercase letter.

DEFINITION. Let X be a real random variable having a probability distribution F_X which is differentiable. Denote the derivative by f_X. Then we call f_X the *probability density function for the r.v.* X. In symbols:

$$f_X(x) = \frac{d}{dx} F_X(x) \tag{3}$$

The values of F_X are probabilities, but the values of f_X are not. Probabilities are found by integrating f_X, for example:

$$P\{a \le X \le b\} = \int_a^b f_X(x)\,dx \tag{4}$$

It follows directly from (2) that the distribution function F_X for any r.v. X possesses the following four properties:

1. F_X is nondecreasing: $a < b$ implies $F_X(a) \le F_X(b)$
2. $\lim_{x \to +\infty} F_X(x) = 1$
3. $\lim_{x \to -\infty} F_X(x) = 0$
4. F_X is continuous from the right, that is, at any discontinuity F_X assumes the upper value.

If F_X is piecewise constant, that is, a staircase function consisting of only finite jumps and constant segments, then X is called a discrete r.v. If F_X has no discontinuities whatsoever, then X is called a *continuous* r.v. A general r.v. is sometimes called *mixed.*

Strictly speaking, only continuous r.v.'s with F_X differentiable can possess density functions, although by resorting to the use of δ functions, which is common in engineering practice, even a discrete r.v. can be assigned a density.

Suppose X is a discrete r.v. which assumes only a finite set of possible values $a_1, a_2, \ldots, a_n$, with respective probabilities $p_1, p_2, \ldots, p_n$. Intuitively, we may say that X can be expected to have value a_k a fraction p_k of the time. If we make many different observations of X and average the results, then as the number of observations becomes infinite the sample average will approach the number

$$\mu = \sum_{k=1}^{n} a_k p_k \tag{5}$$

In (5) we have written μ as a sum over the *range of X*, that is, the set of values assumed by X. Conceptually, it is valuable to realize that this same quantity could also be computed by a sum over the sample space Ω, specifically

$$\mu = \sum_{\omega \in \Omega} X(\omega) P\{\omega \in \Omega: X(\omega) = a_k\} \tag{6}$$

The summation in (6) is accomplished by partitioning Ω into disjoint subsets $A_1, A_2, \ldots, A_n$, such that for each k, A_k is the set of ω points for which $X(\omega)$ assumes the same value a_k.

When X is a continuous r.v., the definition (5) generalizes to

$$\mu = \int_{-\infty}^{\infty} x f_X(x)\, dx \tag{7}$$

The expression (6) generalizes into the Lebesgue integral, as defined in measure theory. A discussion of that is beyond the scope of this book.

The quantity given by (5), (6), or (7) is called the *mean* or *expected value* of X. In rigorous treatments, the most satisfactory way of introducing the expected value operator is to base it on a precise version of (6), which we have here written in a symbolic form to try to suggest the underlying concept.

Since we will mainly be concerned with r.v.'s possessing density functions, we will henceforth take (7) as the definition of the mean, without further comment.

Higher moments are defined analogously, whenever the integrals exist:

$$\mu_n = \int_{-\infty}^{\infty} x^n f_X(x)\, dx \tag{8}$$

When considerable work has to be done involving moments, it is useful to make use of the properties of the *characteristic function* $M(u)$, which is just the Fourier transform of the density:

$$M(u) = \int_{-\infty}^{\infty} e^{iux} f(x)\, dx \tag{9}$$

When the moment μ_n exits, it may be found by the formula

$$\mu_n = (-i)^n \frac{d^n}{du^n} M(u)\bigg|_{u=0} \tag{10}$$

If the characteristic function is known, then the density may be recovered by taking the inverse Fourier transform:

$$f(x) = \frac{1}{2\pi}\int_{-\infty}^{\infty} e^{-iux}M(u)\,du \tag{11}$$

Be careful to note that in the above definitions, the word "inverse," the factor $1/2\pi$, and the minus sign in the exponent have been permuted with respect to the way they are commonly arranged in defining Fourier transforms of functions of time. In dealing with such permutations, the thing that always remains unchanged is the fact that

$$f(x) = \int_{-\infty}^{\infty} \delta(x - x')f(x')\,dx' \tag{12}$$

Now, the δ function can always be represented by either

$$\delta(x - x') = \frac{1}{2\pi}\int_{-\infty}^{\infty} e^{iu(x-x')}\,du \tag{13}$$

or by

$$\delta(x - x') = \frac{1}{2\pi}\int_{-\infty}^{\infty} e^{-iu(x-x')}\,du \tag{14}$$

Depending upon what is called the forward transform and what is called the inverse, it must always be true that either

$$f = \mathscr{F}^{-1}[\mathscr{F}[f]] \tag{15}$$

or

$$f = \mathscr{F}[\mathscr{F}^{-1}[f]] \tag{16}$$

Whichever applies, (15) or (16) must reduce to (12) when the appropriate representation of the δ function is used from (13) or (14).

Fourier transforms occur in this text not only in connection with probability density and characteristic functions, but also in connection with autocovariance and, to be defined later on, power spectral density functions. Since the definitions of these objects do vary from one textbook to another, it is hoped that the above discussion will help dispel some of the resultant confusion. Any variation in the definition is permissible as long as one remains consistent with their own definition and with the above principles.

A new r.v. Y can be generated from an existing r.v. X by making Y a function of X:

$$Y = g(X) \tag{17}$$

Applying the definition of the mean of Y leads us to define, in general, the *expected value of the function* $g(X)$ by

$$E[g(X)] = \int_{-\infty}^{\infty} g(x) f_X(x)\, dx \tag{18}$$

whenever the integral exists.

It is also possible to have several r.v.'s $X_1, X_2, \ldots, X_n$ defined on the same underlying probability trio $(\Omega, \mathscr{A}, P)$, where there is no functional relationship like (17) connecting one r.v. to another. To handle this situation, one uses the *joint cumulative distribution function* $F_{X_1, X_2, \ldots, X_n}(a_1, a_2, \ldots, a_n)$, defined by

$$F_{X_1, X_2, \ldots, X_n}(a_1, a_2, \ldots, a_n)$$
$$= P\{-\infty < X_1 \le a_1, -\infty < X_2 \le a_2, \ldots, -\infty < X_n \le a_n\} \tag{19}$$

If this function is jointly differentiable with respect to all of its arguments, then the *joint density function* $f_{X_1, X_2, \ldots, X_n}(x_1, x_2, \ldots, x_n)$ is defined by

$$f_{X_1, X_2, \ldots, X_n}(x_1, x_2, \ldots, x_n)$$
$$= \frac{\partial^n}{\partial x_1, \partial x_2, \ldots, \partial x_n} F_{X_1, X_2, \ldots, X_n}(x_1, x_2, \ldots, x_n) \tag{20}$$

In this case it is usually expedient to introduce the vector-valued random variable

$$\mathbf{X} = \begin{bmatrix} X_1 \\ X_2 \\ \vdots \\ X_n \end{bmatrix} \tag{21}$$

When this is done, the joint density defined in (20) might be denoted simply $f_{\mathbf{X}}(\mathbf{x})$.

Most of the features of one-dimensional densities can be extended in an obvious way to joint densities. For example, by use of the n-dimensional Fourier transform, a joint characteristic function is defined as a generalization of (9).

Independence and Conditional Probability

Given a probability trio $(\Omega, \mathscr{A}, P)$, let A and B be two members of $\mathscr{A}$. If the measure P assigns probabilities in such a way that

$$P(A \cap B) = P(A)P(B) \tag{22}$$

then we say that the events A and B are *independent*.

Whether A and B are independent or not, if $P(B) > 0$ it is customary to define the ratio

$$\frac{P(A \cap B)}{P(B)} = P(A|B) \tag{23}$$

and $P(A|B)$ is called the *conditional probability of A given B*. In terms of it, the condition (22) for independence may be written

$$P(A|B) = P(A) \tag{24}$$

which says, knowledge of whether or not the event B has occurred has no influence upon the probability that event A occurs.

Suppose X and Y are r.v.'s defined on the same trio $(\Omega, \mathscr{A}, P)$. Assume they are continuous r.v.'s, and let their joint density be $f_{XY}(x, y)$. The two one-dimensional densities for each r.v. considered by itself, denoted respectively $f_X(x)$ and $f_Y(y)$, are called *marginal* densities. They can each be found by *marginal integration*:

$$f_X(x) = \int_{-\infty}^{\infty} f_{XY}(x, y)\, dy \tag{25}$$

$$f_Y(y) = \int_{-\infty}^{\infty} f_{XY}(x, y)\, dx \tag{26}$$

Two r.v.'s that possess a joint density function are *independent if and only if*

$$f_{XY}(x, y) = f_X(x) f_Y(y) \tag{27}$$

The ratio

$$\frac{f_{XY}(x, y)}{f_Y(y)} = f_{X|Y}(x|y) \tag{28}$$

is called the *conditional density for X, given that $Y = y$*.

For example if a and b are two real numbers, then

$$\int_a^b f_{X|Y}(x|y)\, dx = P(\{a \le X \le b\}|\{Y = y\}) \tag{29}$$

If Y is a continuous r.v., then the event $\{Y = y\}$ has probability zero. The conditional probability on the right-hand side of (29) therefore cannot be defined by a simple straightforward application of the definition (23). The conditional probability in (29) can, nevertheless, be defined in a way that is totally satisfactory from a rigorous standpoint, but it requires use of a technical device from measure theory, called a sigma-field, which is beyond the scope of this text.

Let $X_1, X_2, \ldots, X_n$ all be continuous r.v.'s defined on the same trio $(\Omega, \mathscr{A}, P)$, and suppose they have a joint density $f_{X_1 X_2 \cdots X_n}(x_1, x_2, \ldots, x_n)$. Let m be an integer such that $1 < m < n$. The conditional density for $X_{m+1}, X_{m+2}, \ldots, X_n$, given $X_1, X_2, \ldots, X_m$, denoted $f_{X_{m+1},\ldots,X_n|X_1,\ldots,X_m}(x_{m+1},\ldots,x_n|x_1,\ldots,x_m)$ is defined as the ratio

$$f_{X_{m+1},\ldots,X_n|X_1,\ldots,X_m}(x_{m+1},\ldots,x_n|x_1,\ldots,x_m) = \frac{f_{X_1,X_2,\ldots,X_n}(x_1,x_2,\ldots,x_n)}{f_{X_1,X_2,\ldots,X_m}(x_1,x_2,\ldots,x_m)} \tag{30}$$

The denominator of (30) is found by marginal integration:

$$f_{X_1,X_2,\ldots,X_m}(x_1,x_2,\ldots,x_m) = \underbrace{\int_{-\infty}^{\infty} \cdots \int_{-\infty}^{\infty}}_{n-m} f_{X_1,X_2,\ldots,X_n}(x_1,x_2,\ldots,x_n)\, dx_{m+1},\ldots, dx_n \tag{31}$$

The Hilbert Space of Second-Order Random Variables

A *Hilbert Space* is a vector space equipped with an inner product and a norm that is derived from the inner product. Finite-dimensional Hilbert space is just an abstraction and generalization of finite-dimensional Euclidean space. Infinite-dimensional Hilbert space is the extension of this concept to an infinite number of dimensions. In that case the definition must also be expanded to include the attribute "complete." "Complete" means that every infinite sequence of vectors drawn from the space, which is Cauchy in the sense of the norm, converges to a limit vector that also belongs to the space.

We will discuss the finite-dimensional case first. Suppose we have a set of n random variables $X_1, X_2, \ldots, X_n$, each of which has finite second moment:

$$EX_k^2 < \infty, \qquad k = 1, 2, \ldots, n \tag{32}$$

An r.v. X_k that obeys (32) is called a *second-order* r.v.

Our first task is to introduce the concepts of linear independence and statistical independence.

DEFINITION. The set of second-order r.v.'s $X_1, X_2, \ldots, X_n$ is called *linearly independent* if and only if the equation

$$\begin{aligned} c_1X_1 + c_2X_2 + \cdots + c_nX_n &= 0 \\ \text{implies} \quad c_1 = c_2 = \cdots = c_n &= 0 \end{aligned} \tag{33}$$

Let $F_{X_1, X_2, \ldots, X_n}(x_1, x_2, \ldots, x_n)$ be the joint cumulative distribution of $X_1, X_2, \ldots, X_n$, that is,

$$F_{X_1, X_2, \ldots, X_n}(x_1, x_2, \ldots, x_n) = P\{-\infty < X_k \le x_k,\ k = 1, 2, \ldots, n\}$$

Let $F_{X_k}(x_k)$, $k = 1, 2, \ldots, n$ be the marginal cumulative distribution function for each r.v. X_k, $k = 1, 2, \ldots, n$.

DEFINITION. The set of r.v.'s $X_1, X_2, \ldots, X_n$ is called *statistically independent* if and only if

$$F_{X_1, X_2, \ldots, X_n}(x_1, x_2, \ldots, x_n) = \prod_{k=1}^{n} F_{X_k}(x_k) \tag{34}$$

This is the usual definition of mutual independence, generalizing (27). The r.v.'s $X_1, \ldots, X_n$ need not have finite second moments in order for this definition to be usable. Also, there is no assumption that the means $E[X_n]$ are zero. However, if they all have both zero mean and finite second moment, then statistical independence implies linear independence, but not conversely.

DEFINITION. Let X_j and X_k be second-order r.v.'s. Their *inner product*, denoted $\langle X_j, X_k \rangle$, is defined as

$$\langle X_j, X_k \rangle = E[X_j X_k] \tag{35}$$

DEFINITION. Let X be a second-order r.v. Its *norm*, denoted $\|X\|$, is defined as

$$\|X\| = \sqrt{E[X^2]} \tag{36}$$

Now let $X_1, X_2, \ldots, X_n$ be any set of n linearly independent second-order r.v.'s Consider the set of all possible liner combinations of these r.v.'s, that is, all other r.v.'s of the form

$$Y = \sum_{k=1}^{n} c_k X_k \tag{37}$$

We then have

$$\begin{aligned} Y^2 &= \left(\sum_{k=1}^{n} c_k X_k \right)^2 \\ &= \sum_{k=1}^{n} \sum_{j=1}^{n} c_k c_j X_k X_j \end{aligned} \tag{38}$$

where we have written the product of two single sums as a double summation by changing the dummy index.

One of the problems at the end of the chapter is to prove the Schwarz inequality

$$|\langle X, Y \rangle| \leq \|X\| \, \|Y\| \tag{39}$$

Taking the expected value of both sides of (38) yields

$$\begin{aligned} \|Y\|^2 = E[Y^2] &= \sum_{h=1}^{n} \sum_{j=1}^{n} c_k c_j E[X_k X_j] \\ &= \sum_{k-1}^{n} \sum_{j=1}^{n} c_k c_j \langle X_k, X_j \rangle \\ &\leq \sum_{k=1}^{n} \sum_{j=1}^{n} |c_k| \, |c_j| \, |\langle X_k, X_j \rangle| \\ &\leq \sum_{k=1}^{n} \sum_{j=1}^{n} |c_k| \, |c_j| \, \|X_k\| \, \|X_j\| \\ &= \left(\sum_{k=1}^{n} |c_k| \, \|X_k\| \right)^2 \end{aligned} \tag{40}$$

Thus $\|Y\|^2 \leq (\sum_{k=1}^{n} |c_k| \, \|X_k\|)^2 < \infty$ provided $|c_k| < \infty$, $k = 1, 2, \ldots, n$, so every such r.v. of the form (37) is a second-order r.v. The set of all such r.v.'s Y is our Hilbert space.

We may construct an orthogonal basis for the space $V_1, V_2, \ldots, V_n$ by applying the Gram–Schmidt procedure to $X_1, X_2, \ldots, X_n$:

$$V_1 = X_1$$
$$V_2 = X_2 - \frac{\langle X_2, V_1\rangle}{\langle V_1, V_1\rangle} V_1 \tag{41a}$$

and for arbitrary k,

$$V_k = X_k - \sum_{j=1}^{k-1} \frac{\langle X_k, V_j\rangle}{\langle V_j, V_j\rangle} V_j \tag{41b}$$

If the original set $\{X_1, X_2, \ldots, X_n\}$ were not in fact all linearly independent, then the above procedure will simply return zero for the corresponding V_k, whenever X_k is not linearly independent of $\{X_1, \ldots, X_{k-1}\}$. In that case, merely continue with the procedure, deleting V_k. When finished, the resulting set $\{V_1, V_2, \ldots, V_m\}$ for some $m < n$ will be an orthogonal basis, and m will be the dimension of the space spanned by $\{X_1, X_2, \ldots, X_n\}$.

The same procedure can be used in the case of infinitely many elements $X_1, X_2, \ldots,$ and in principle could be used to determine whether the HIlbert space is finite-dimensional or infinite-dimensional. The possibility that we may be dealing with an *infinite* number of r.v.'s all defined on the same probability trio $(\Omega, \mathscr{A}, P)$, which are mutually linearly independent, will be very important to us for the remainder of this book, because it is a fundamental concept in the theory of *random processes*, to which we now turn.

Random Processes

Having raised the possibility of an infinite family of r.v.'s all defined on the same sample space, we now formalize the concept.

DEFINITION. A *random process* (equivalently: *stochastic process*) is a family $\{X_t: t \in T\}$ of random variables, all defined on the same probability trio $(\Omega, \mathscr{A}, P)$. The set T is the *parameter set* of the random process.

In this book, all of our random processes will either be real valued, complex valued, or vector valued (vector in R^n). If the parameter set T is the set of integers or a subset thereof, the process is called a *discrete parameter* process. If T is a subset of the real line, then the process is called a *continuous parameter* process.

In general, other possibilities exist for the space in which the r.v.'s take values (sometimes called the *state space* of the process) and for the parameter set. The ones mentioned above are the only ones used in this book.

In Chapters 7 and 10 we will discuss a particular category of random processes known as Markov processes. It is appropriate to provide the relevant definition here.

DEFINITION. The random process $\{X_t: t \in T\}$ is called a *Markov process* provided the following circumstances hold:

Let S be any subset of the state space. Let t_f (the future time) and t_p (the present time) be any two elements of T with $t_f > t_p$. Let Q (the set of past times) be any subset of T containing t_p such that, for every $t \in Q$, $t_p \geq t$. Then

$$P\{X_{t_f} \in S | X_t, t \in Q\} = P\{X_{t_f} \in S | X_{t_p}\} \tag{42}$$

In words, this definition says that a Markov process is one having the property that, given the present state of the process, the future becomes conditionally independent of the past. To illustrate this further, suppose $\{X_t: t \in T\}$ is a Markov process such that all of the r.v.'s X_t are continuous r.v.'s. Let $t_1 < t_2 < \cdots < t_{n-1} < t_n$ be a set of points in T. Then we may consider the joint density function $f_{X_{t_1}, X_{t_2}, \ldots, X_{t_n}}(x_1, x_2, \ldots, x_n)$ and the associated conditional density function (with $1 < m < n$)

$$f_{X_{t_{m+1}}, X_{t_{m+2}}, \ldots, X_{t_n} | X_{t_1}, X_{t_2}, \ldots, X_{t_m}}(x_{m+1}, x_{m+2}, \ldots, x_n | x_1, x_2, \ldots, x_m)$$

as defined in (30). Let t_m play the role of the present time t_p in the preceding definition. Consider $t_{m+1}, \ldots, t_n$ as future times, and $t_1, \ldots, t_{m-1}$ as past times. Then, for *any* set $\{t_1, t_2, \ldots, t_n\}$ chosen from T with the preceding properties, the Markov nature of $\{X_t: t \in T\}$ means that the following equation is *identically true*:

$$\begin{aligned} &f_{X_{t_{m+1}}, \ldots, X_{t_n} | X_{t_1}, \ldots, X_{t_m}}(x_{m+1}, \ldots, x_n | x_1, \ldots, x_m) \\ &\quad = f_{X_{t_{m+1}}, \ldots, X_{t_n} | X_{t_m}}(x_{m+1}, \ldots, x_n | x_m) \end{aligned} \tag{43}$$

That completes our consideration of Markov processes in this chapter. We return to the discussion of general random processes.

Since the qualification "$t \in T$" is always understood, henceforth in this book we shall simply write a random process as $\{X_t\}$. Again let

$t_1 < t_2 < \cdots < t_n$ be an ordered set of parameter points (usually the parameter will be interpreted as time). Consider the random vector $\mathbf{X}$ of the values $X_{t_1}, X_{t_2}, \ldots, X_{t_n}$, as in (21), and denote the joint density function by $f_{\mathbf{X}}(\mathbf{x})$.

Let $\mathbf{C}$ be an $n \times n$ matrix of real numbers, which is symmetric and positive definite. Let $\boldsymbol{\mu}$ be an n-vector of real numbers. We will provide a fuller discussion of Gaussian distributions in Chapters 2 and 3, but it is appropriate to introduce the following definition now:

DEFINITION. The random process $\{X_t\}$ is called a *Gaussian* process provided that for any selection of the ordered set $\{t_1, t_2, \ldots, t_n\}$ from T, for any integer n, there exists a positive definite $n \times n$ matrix $\mathbf{C}$ and an n-vector $\boldsymbol{\mu}$ such that the joint density function for the random vector $\mathbf{X}$ pertinent to these time points is given by

$$f_{\mathbf{X}}(\mathbf{x}) = \frac{|\mathbf{C}|^{-1/2}}{(2\pi)^{n/2}} \exp\{-\tfrac{1}{2}(\mathbf{x}-\boldsymbol{\mu})^T\mathbf{C}^{-1}(\mathbf{x}-\boldsymbol{\mu})\} \tag{44}$$

Much of the remainder of this book is devoted to the consideration of Gaussian processes. In (44), the notation $|\mathbf{C}|$ means the determinant of the matrix $\mathbf{C}$.

Two very important entities associated with any Gaussian process $\{X_t\}$ are its *mean* μ_t,

$$\mu_t = E\{X_t\} \tag{45}$$

and its *autocovariance* c_{ts},

$$c_{ts} = E\{[X_t - \mu_t][X_s - \mu_s]\} \tag{46}$$

both defined for all t, s in T. Conversely, we show later that knowledge of these two entities is sufficient to characterize completely a Gaussian process.

Be careful not to jump to the conclusion that, just because the mean μ_t and covariance c_{ts} are given for some process $\{X_t\}$, it necessarily follows that $\{X_t\}$ is Gaussian. This conclusion is false. There are many possible distribution laws having the same first two moments as a particular Gaussian distribution, but for which the higher moments are entirely different. As a specific example, the distribution could be bimodal. There is no way to tell whether a distribution is unimodal, bimodal, or multimodal by looking only at the first two moments.

In the applications for the material presented in this book, one is usually dealing with a physical situation in which there is an element of randomness and where important features of the situation are investigated by examining a collection of measurements that emerge sequentially in time. That is the sort of situation one would attempt to model by means of a random process. After constructing such a model one would hope that mathematical analysis of the model will lead one to insights that are valid and pertinent to the actual physical situation.

In the selection and construction of such a model, there is always the question of how detailed and explicit to make it. In the present context, this question would be relevant to the selection of an appropriate model when all one knows about some process $\{X_t\}$ are data that are equivalent to the first two moments, or equivalently, to the mean μ_t and the autocovariance c_{ts}. The decision that has to be made in that case is whether to commit oneself to a specific probability distribution law or whether to leave that feature open.

The disadvantage of choosing a model based on an explicitly specified distribution (e.g., Gaussian) is that one has made the model more detailed and explicit than the situation, or the data available, actually warrants. In turn, this leads one into a false sense of confidence to make inferences and extrapolate results far beyond what is justified by the knowledge available. This is the peril of overspecificity.

On the other hand, if one chooses to duck the issue and assume no specific probability law, then the only calculations that can ever be made are those based specifically (in the case under consideration) on the first two moments. Means and covariances of various r.v.'s can be calculated, but it is never possible to calculate probabilities of events.

If the only operations ever performed on the processes under consideration are linear, then it turns out that only the pertinent mean and covariance functions ever need be considered. That is, the mean, autocovariance of the output, and cross-covariance between output and input of a linear system can be calculated knowing only the mean and autocovariance of the input and the transfer function of the system. A body of theory exists for carrying out precisely such calculations for second-order processes in linear systems. In this way a stochastic system can be analyzed using only deterministic quantities. For many years, this was considered to be a great advantage.

There is a recent development that tends to reverse the situation: the widespread popularity of computer simulation. Rather than just using the computer to carry out the above-mentioned calculations based on second-order theory, the incredibly high speed of modern computers makes it feasible both technically and economically to do so-called Monte Carlo simulations of stochastic systems. In such a simulation, the computer generates an ensemble of waveforms with the appropriate statistical characteristics. These waveforms are applied, one by one, as inputs to the system, and the resulting outputs are recorded. Statistical analysis of the resulting ensemble of outputs then permits one to make whatever calculations and inferences are appropriate to the system under consideration.

Because Monte Carlo simulation can be performed just as readily for time-varying and nonlinear systems as for time-invariant linear systems, this approach is steadily gaining wider acceptance and greater favor. The reason the use of Monte Carlo simulations reverse the preference for second-order models over specifically Gaussian models is that, in setting up the simulation, one has to adopt some specific distribution for the random numbers being generated. There are more compelling reasons to choose the Gaussian distribution than any other with the same second-order statistics.

It is quite likely that any contemporary serious worker in applied stochastic processes will become involved in computer simulations. In this work, he will find it very handy to know some techniques for working with actual Gaussian distributions, beyond the methods of second-order theory. For this reason, this book includes considerable discussion of certain features of Gaussian distributions that are likely to be useful in running computer simulations. In particular, the next three chapters focus on this material.

Starting with the simple one-dimensional Gaussian distribution in Chapter 2, we cover the multidimensional distribution in Chapter 3 and move toward random process theory by discussing sequences of finite length in Chapter 4. Finally in Chapter 5 we present the classical second-order theory for discrete time sequences, and move to continuous-time processes in Chapter 6. Chapters 7, 8, and 9 cover more advanced subjects involving continuous-time stochastic processes. Chapters 10 and 11 cover some advanced subjects involving discrete-time stochastic processes. After having completed the study of Chapters 1–6, the sets $\{7, 8, 9\}$ and $\{10, 11\}$ are independent of each other. Either set may be studied on its own, at the students' (or the instructor's) convenience and discretion.

Another recurring theme throughout this book is triangular factorization of covariance matrices, which finds its parallel in continuous time in the spectral factorization technique. In order to attempt to dispel the mystery of this topic, it is introduced in the next chapter in the familiar procedure of completing the square. Appendix 1 gives the basic pertinent theorem. The idea reaches another culmination in the final chapter of the book, where it is used to derive the Kalman filter via the concept of the innovations process. In the discussion of the innovations process in Chapter 11, we will again meet the Gram–Schmidt orthogonalization procedure in the Hilbert space of second-order random variables.

This book not only moves upward and outward to provide the reader with an ongoing confrontation with new topics, but also periodically revisits topics already discussed, showing how they reappear in a new guise. To some extent, the organization of the book therefore resembles a spiral. We hope the reader enjoys the journey.

Problems

1. The sample space S has three elements: $S = \{s_1, s_2, s_3\}$. The set function $Q(\cdot)$ assigns numbers as follows:

$$Q(\{s_1\}) = \tfrac{1}{4} \qquad Q(\{s_3\}) = \tfrac{1}{4}$$

$$Q(\{s_2\}) = \tfrac{1}{4} \qquad Q(S) = 1$$

Is $Q(\cdot)$ a probability measure? If not, what conditions must be changed to make it one?

2. For every subset A of the real line, let $N(A)$ be the set function whose value at A is equal to the number of points in A which are positive integers. For example, if $A_0 = \{x\colon 6\tfrac{1}{2} \le x \le 7\tfrac{1}{2}\}$, then $N(A_0) = 1$.

Now let

$$A_1 = \{x\colon x \text{ is a multiple of 3 and } x \le 50\}$$

$$A_2 = \{x\colon x \text{ is a multiple of 7 and } x \le 50\}$$

a. Find $N(A_1)$, $N(A_2)$, $N(A_1 \cup A_2)$, $N(A_1 \cap A_2)$.

b. Verify that $N(A_1 \cup A_2) = N(A_1) + N(A_2) - N(A_1 \cap A_2)$.

3. A bag contains seven red balls and three green balls. A box contains five red balls and six green ones. Three balls are selected at random from the bag and are transferred to the box, after which a ball is selected at random from the box.

a. What is the probability that the ball drawn from the box is red?

b. Given that the ball from the box is red, what is the conditional probability that two or more of the balls transferred were red?

4. The bivariate r.v. (X, Y) has the joint density function

$$f_{XY}(x, y) = \begin{cases} \dfrac{x^2 - y^2}{8} e^{-x}, & 0 \le x \le \infty,\ -x \le y \le x \\ 0, & \text{otherwise} \end{cases}$$

Find the marginal densities $f_X(x)$ and $f_Y(y)$, and the conditional density $f_{Y|X}(y|X = x)$.

5. Let X be a Gaussian random variable having the probability density function

$$f_X(u) = (2\pi)^{-1/2}\exp\left(-\frac{u^2}{2}\right)$$

The random variable Y is defined as $Y = X^3$. Determine and plot the probability density function for Y.

6*. The stochastic process $X(t)$ is defined by

$$X(t) = \sin(at + B)$$

where a is constant and B is a random variable uniformly distributed on the interval $[0, 2\pi]$. Find the cumulative distribution function

$$F_X(u) = P\{-\infty < X(t) \le u\}$$

and the probability density function

$$f_X(u) = \frac{d}{du}F_X(u)$$

7*. The stochastic process $X(t)$ is defined by

$$X(t) = e^{tA}, \qquad 0 \le t \le \infty$$

where A is a random variable with probability density function $f_A(a)$. Find the distribution function

$$F_X(u) = P\{-\infty < X(t) \le u\}$$

and the density

$$f_X(u) = \frac{d}{du}F_X(u)$$

8. Urn #1 and urn #2 each initially contain six red, four white, and eight blue balls. At each step, one ball is selected at random from each urn, and the two balls interchange urns.

**Note:* In Problems 6 and 7, the distribution and density functions will also depend on the time t.

At time n, let the random variable X_n be the number of white balls in urn #1. Is the random sequence $X_0, X_1, X_2, \ldots, X_n, \ldots,$ a Markov chain?

If you say yes, then determine the transition function $P(x, y)$ defined as $P(x, y) = P\{X_k = x | X_{k-1} = y\}$.

If you say no, explain why it's not Markov.

9. Suppose the procedure in Problem 8 is modified as follows: At each step, one ball is still selected at random from each urn. However, if it is red, then it is replaced in the *same* urn from which it was drawn. If it is blue or white, then it is placed in the opposite urn.

Repeat Problem 8 for this situation.

10. Let $\mathscr{H}$ be a Hilbert space, with inner product $\langle \cdot, \cdot \rangle$ and norm $\|\cdot\|$. Prove the *Schwarz inequality*: for any vectors x, y in $\mathscr{H}$, it holds that

$$|\langle x, y \rangle| \leq \|x\| \cdot \|y\|$$

Hint: Start from the fact that for all choices of scalars α, β, it holds that

$$\langle (\alpha x + \beta y), (\alpha x + \beta y) \rangle \geq 0$$

and choose α and β suitably.

Chapter 2

The Gaussian Distribution in One and Two Dimensions

The One-Dimensional Gaussian Distribution

A real-valued random variable X is said to possess a *Gaussian distribution* if, for real numbers a, b with $a < b$ it holds that

$$P\{a \leq X < b\} = \frac{1}{\sqrt{2\pi\sigma^2}} \int_a^b e^{-(x-\mu)^2/2\sigma^2}\, dx \tag{1}$$

In this expression, μ and σ are parameters of the distribution. By direct calculation, one finds that

$$E\{X\} = \mu$$

$$E\{(X-\mu)^2\} = \sigma^2$$

so that μ is the mean and σ^2 is the variance of the random variable X.

The integral in (1) cannot be evaluated analytically, so in order to do numerical calculations, one must resort to tables.

The function under the integral sign in (1), specifically,

$$f(x) = \frac{1}{\sqrt{2\pi\sigma^2}} e^{-(x-\mu)^2/2\sigma^2} \tag{2}$$

is called the *Gaussian probability density function*. We assume that its principal features are already familiar to the student.

One important property is that it is necessary to construct tables only for the normalized variable

$$u = \frac{x - \mu}{\sigma} \tag{3}$$

and the corresponding density

$$\phi(u) = \frac{1}{\sqrt{2\pi}} e^{-u^2/2} \tag{4}$$

because the properties of the general density (2) can be recovered by use of the transformation (3).

The function $\phi(u)$ is plotted in Figure 2.1. One particular property is symmetry, that is, $\phi(u) = \phi(-u)$, from which it follows that

$$\int_0^{\infty} \phi(u)\, du = \tfrac{1}{2} \tag{5}$$

The function usually given in tables and used for evaluating integrals such as (1) is

$$\Phi(y) = \int_0^{y} \phi(u)\, du \tag{6}$$

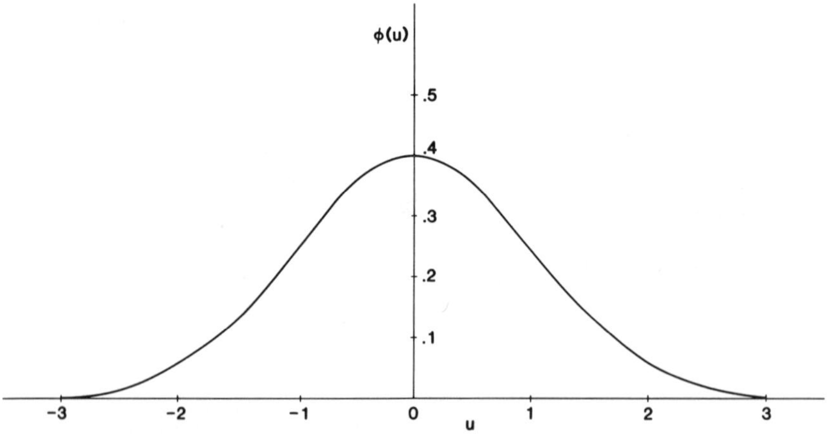

FIGURE 2.1. Plot of the standard Gaussian density function.

For example, to evaluate (1), make the change of variables (3). Then we have

$$P\{a \leq X < b\} = \frac{1}{\sqrt{2\pi}} \int_{(a-\mu)/\sigma}^{(b-\mu)/\sigma} e^{-u^2/2}\, du \tag{7}$$

If $b > a > 0$, then we have at once

$$P\{a \leq X < b\} = \Phi\left(\frac{b-\mu}{\sigma}\right) - \Phi\left(\frac{a-\mu}{\sigma}\right) \tag{8}$$

The clever student can figure out how to handle other cases for himself.

Many of the random variables to be considered in this course will have Gaussian distributions. Since we will consider several such random variables at the same time, we must now discuss joint Gaussian distributions.

The Two-Dimensional Gaussian Distribution

A pair of random variables X_1, X_2 is said to be jointly Gaussian if, for real numbers a_1, b_1, a_2, b_2 with $a_1 < b_1$ and $a_2 < b_2$, it holds that

$$P\{a_1 \leq X_1 < b_1, a_2 \leq X_2 < b_2\}$$

$$= \frac{|\mathbf{C}|^{-1/2}}{2\pi} \int_{a_1}^{b_1} \int_{a_2}^{b_2} \exp\left\{ -\frac{1}{2}[(x_1 - \mu_1)(x_2 - \mu_2)] \times \begin{bmatrix} s_{11} & s_{12} \\ s_{21} & s_{22} \end{bmatrix} \begin{bmatrix} (x_1 - \mu_1) \\ (x_2 - \mu_2) \end{bmatrix} \right\} dx_1\, dx_2 \tag{9}$$

In this case, the parameters of the distribution are five numbers $\mu_1, \mu_2, c_{11}, c_{12}, c_{22}$. The matrix $\mathbf{C}$ is defined as

$$\mathbf{C} = \begin{bmatrix} c_{11} & c_{12} \\ c_{12} & c_{22} \end{bmatrix} \tag{10}$$

Note that this matrix is symmetric: $c_{21} = c_{12}$.

Its inverse is denoted by $\mathbf{S}$:

$$\mathbf{C}^{-1} = \mathbf{S} = \begin{bmatrix} s_{11} & s_{12} \\ s_{21} & s_{22} \end{bmatrix} \tag{11}$$

Since $\mathbf{C}$ is symmetric, then necessarily also $s_{21} = s_{12}$.

The symbol $|\mathbf{C}|$ in front of the integral in (9) stands for the determinant of $\mathbf{C}$, that is,

$$|\mathbf{C}| = c_{11}c_{22} - c_{12}^2$$

By Cramer's rule, we may write (11) in terms of the elements of $\mathbf{C}$:

$$\mathbf{S} = |\mathbf{C}|^{-1}\begin{bmatrix} c_{22} & -c_{12} \\ -c_{12} & c_{11} \end{bmatrix} \tag{12}$$

The function under the integral sign in (9) is the *two-dimensional joint Gaussian probability density function*:

$$f(x_1, x_2) = \frac{|\mathbf{C}|^{-1/2}}{2\pi}\exp\left\{-\frac{1}{2}[(x_1-\mu_1)(x_2-\mu_2)]\begin{bmatrix} s_{11} & s_{12} \\ s_{12} & s_{22} \end{bmatrix}\begin{bmatrix} (x_1-\mu_1) \\ (x_2-\mu_2) \end{bmatrix}\right\} \tag{13}$$

The matrix $\mathbf{C}$ in (10) is called the *covariance matrix*. Its meaning is as follows:

$$\mathbf{C} = E\left\{\begin{bmatrix} (X_1-\mu_1) \\ (X_2-\mu_2) \end{bmatrix}[(X_1-\mu_1)(X_2-\mu_2)]\right\}$$

$$= E\left\{\begin{bmatrix} (X_1-\mu_1)^2 & (X_1-\mu_1)(X_2-\mu_2) \\ (X_2-\mu_2)(X_1-\mu_1) & (X_2-\mu_2)^2 \end{bmatrix}\right\} \tag{14}$$

In (14), we have formed the outer product of the vector $[(X_1-\mu_1)(X_2-\mu_2)]$ with itself to obtain a 2×2 matrix, and then have taken the expected value of this random matrix to obtain the covariance matrix $\mathbf{C}$. This will be a commonly occurring procedure in what follows. From (14) we see that

$$c_{12} = E\{(X_1-\mu_1)(X_2-\mu_2)\} \tag{15}$$

Therefore, $c_{12} = 0$ if and only if the random variables X_1 and X_2 are uncorrelated. In the case of Gaussian random variables, unlike the general case, if X_1 and X_2 are uncorrelated, then they are actually independent.

It is easy to see that this is so. If $c_{12} = 0$, then from (12), $s_{12} = 0$ also. Moreover, $|\mathbf{C}| = c_{11}c_{22}$. Therefore, (13) becomes

$$\begin{aligned}
f(x_1, x_2) &= \frac{1}{2\pi\sqrt{c_{11}c_{22}}} \exp\left\{-\frac{1}{2}\left[\frac{(x_1-\mu_1)^2}{c_{11}} + \frac{(x_2-\mu_2)^2}{c_{22}}\right]\right\} \\
&= \frac{1}{\sqrt{2\pi c_{11}}} \exp\left[-\frac{(x_1-\mu_1)^2}{2c_{11}}\right] \frac{1}{\sqrt{2\pi c_{22}}} \exp\left[-\frac{(x_2-\mu_2)^2}{2c_{22}}\right] \\
&= f_1(x_1) f_2(x_2)
\end{aligned}$$

That is, when $c_{12} = 0$, then the joint density $f(x_1, x_2)$ factors into the product of the two one-dimensional densities $f_1(x_1)$ and $f_2(x_2)$.

Integration Over a Quadrant

Since the effect of $\mu_1 \neq 0$ and $\mu_2 \neq 0$ is simply a translation, let us now assume $\mu_1 = \mu_2 = 0$ to simplify our expressions. Consider the integral

$$I = \int_0^\infty \int_0^\infty f(x_1, x_2)\, dx_1\, dx_2 \tag{16}$$

In the case when $\mu_1 = \mu_2 = 0$ and also $c_{12} = 0$, we have

$$\begin{aligned}
I &= \int_0^\infty \int_0^\infty f(x_1, x_2)\, dx_1\, dx_2 \\
&= \int_0^\infty \int_0^\infty f_1(x_1) f_2(x_2)\, dx_1\, dx_2 \\
&= \left(\int_0^\infty f_1(x_1)\, dx_1\right)\left(\int_0^\infty f_2(x_2)\, dx_2\right) \\
&= \tfrac{1}{2} \cdot \tfrac{1}{2} = \tfrac{1}{4}
\end{aligned} \tag{17}$$

Now let us examine the case when $c_{12} \neq 0$. Since we wish to introduce a specific computational technique, the discussion will proceed most simply if we consider a specific numerical example. Therefore, suppose $\mu_1 = \mu_2 = 0$ and

$$\mathbf{C} = \begin{bmatrix} 5 & 3 \\ 3 & 2 \end{bmatrix} \tag{18}$$

From (12),

$$\mathbf{S} = \begin{bmatrix} 2 & -3 \\ -3 & 5 \end{bmatrix} = \mathbf{C}^{-1} \tag{19}$$

and (13) becomes in this case

$$f(x_1, x_2) = \frac{1}{2\pi}\exp\left\{-\frac{1}{2}[x_1 \quad x_2]\begin{bmatrix} 2 & -3 \\ -3 & 5 \end{bmatrix}\begin{bmatrix} x_1 \\ x_2 \end{bmatrix}\right\}$$

$$= \frac{1}{2\pi}\exp\left(-\frac{2x_1^2 - 6x_1x_2 + 5x_2^2}{2}\right) \tag{20}$$

We wish to evaluate the integral (16), which becomes in this case

$$I = \frac{1}{2\pi}\int_0^\infty \int_0^\infty \exp(-x_1^2 + 3x_1x_2 - \tfrac{5}{2}x_2^2)\, dx_1\, dx_2 \tag{21}$$

The technique we wish to illustrate is called *completing the square*. The multinomial in the exponent may be written

$$-x_1^2 + 3x_1x_2 - \tfrac{5}{2}x_2^2 = -x_1^2 + 3x_1x_2 - \tfrac{9}{4}x_2^2 + \tfrac{9}{4}x_2^2 - \tfrac{5}{2}x_2^2$$

$$= -(x_1 - \tfrac{3}{2}x_2)^2 - \tfrac{1}{4}x_2^2 \tag{22}$$

What we did was to add and subtract the quantity $\frac{9}{4}x_2^2$, so that the expression could be written as the algebraic sum of two perfect squares.

Now make the change of variables

$$y = x_1 - \tfrac{3}{2}x_2$$

$$z = \tfrac{1}{2}x_2$$

This may be written in matrix form as

$$\begin{bmatrix} y \\ z \end{bmatrix} = \begin{bmatrix} 1 & -\frac{3}{2} \\ 0 & \frac{1}{2} \end{bmatrix}\begin{bmatrix} x_1 \\ x_2 \end{bmatrix} \tag{23}$$

The determinant of the matrix, which is the Jacobian of the transformation, equals $\frac{1}{2}$, so $dy\, dz = \frac{1}{2}\, dx_1\, dx_2$. With this change, (21) becomes

$$I = \frac{1}{\pi}\int_0^\infty \left(\int_{-3z}^\infty e^{-y^2}e^{-z^2}\, dy\right) dz$$

$$= \frac{1}{\pi}\int_0^\infty e^{-z^2}\left(\int_{-3z}^\infty e^{-y^2}\, dy\right) dz \tag{24}$$

The lower limit on the inner integral is $-3z$, because from (23), when $x_1 = 0$, then $y = -\frac{3}{2}x_2 = -3z$.

Because of the presence of this lower limit, the double integral in this case does not factor into the product of two one-dimensional integrals. It remains an iterated integral, that is, we must do the integration of y first to obtain a function of z only, and then combine that with the rest of the integrand e^{-z^2}, and perform the integration on z.

Since we have said that this inner integral on y cannot be evaluated analytically, but only numerically by use of tables, it appears that we have reached an analytical impasse. Actually, there is a way out. What we do is to transform to polar coordinates, that is, define

$$\begin{aligned} y &= r\cos\theta \\ z &= r\sin\theta \end{aligned} \tag{25}$$

Solving for r and θ yields

$$\begin{aligned} r &= \sqrt{y^2 + z^2} \\ \theta &= \tan^{-1}\frac{z}{y} \end{aligned} \tag{26}$$

Let us sketch the region in the plane over which the integral (24) is to be evaluated. A little thought will show that it is the shaded region indicated in Figure 2.2. The integration over y covers everything to the right of the

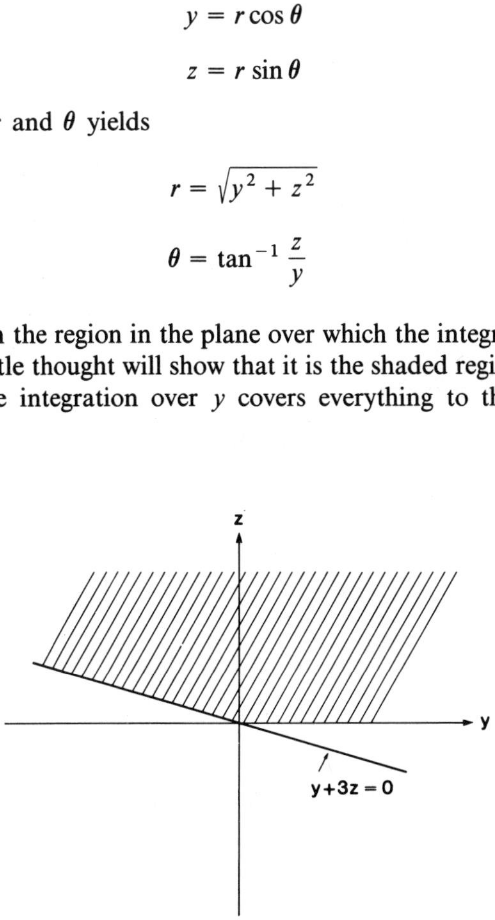

FIGURE 2.2. The region of integration.

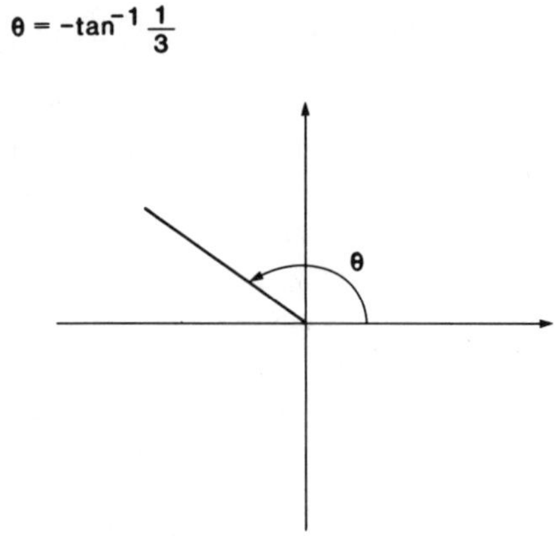

FIGURE 2.3. Polar coordinates.

sloping line whose equation is $y + 3z = 0$. The integration over z covers everything above the y-axis, that is, the line $z = 0$. Combining these conditions, we see the integration is to be carried out over the wedge-shaped shaded region in Figure 2.2. This figure is redrawn in polar coordinates in Figure 2.3. We see that the limits on the integral should be $0 \leq r < \infty$, $0 \leq \theta < -\tan^{-1}\frac{1}{3}$. The angle whose tangent is $-\frac{1}{3}$ is about 161° 34′, or 2.82 radians.

In polar coordinates the integral (24) is

$$I = \frac{1}{\pi}\int_0^{\infty}\int_{-3z}^{\infty} e^{-(y^2+z^2)}\,dy\,dz$$

$$= \frac{1}{\pi}\int_0^{2.82}\int_0^{\infty} e^{-r^2} r\,dr\,d\theta \qquad (27)$$

Now make the change of variable $q = r^2$, $dq = 2r\,dr$. In terms of q and θ we have

$$I = \frac{1}{2\pi}\int_0^{2.82}\left(\int_0^{\infty} e^{-q}\,dq\right)d\theta \qquad (28)$$

Now

$$\int_0^{\infty} e^{-q}\,dq = 1$$

and

$$\int_0^{2.82} d\theta = 2.82$$

so finally

$$\boxed{I = \frac{2.82}{2\pi} = .4488} \qquad (29)$$

Summarizing now, we have been considering the integral (16) for the case $\mu_1 = \mu_2 = 0$. In (12) we showed that when X_1 and X_2 are uncorrelated, then regardless of the values of c_{11} and c_{22}, the integral I always equals $\frac{1}{4}$. We then considered a specific numerical example in which X_1 and X_2 are correlated, given by (21). One of our purposes was to illustrate the computational methods commonly used to evaluate such an integral. Another purpose was to arrive at the conclusion that correlation can drastically affect the value of the integral, and in that case it is not true that the value of the integral over one quadrant is $\frac{1}{4}$. Of course, we *always* have

$$\int_{-\infty}^{\infty} \int_{-\infty}^{\infty} f(x_1, x_2)\, dx_1\, dx_2 = 1 \qquad (30)$$

because f is a probability density function.

Problems

1. Let $f_{XY}(x, y)$ be the density function representing a joint Gaussian pair of random variables, with mean vector zero and covariance matrix

$$\mathbf{C} = \begin{bmatrix} 3 & 2 \\ 2 & 2 \end{bmatrix}$$

a. Write out $f_{XY}(x, y)$ explicitly.

b. Compute $\int_0^\infty \int_0^\infty f_{XY}(x, y)\, dx\, dy$.

2. Repeat problem 1 for

$$\mathbf{C} = \begin{bmatrix} 6 & 4 \\ 4 & 3 \end{bmatrix}$$

3. If $f_{XY}(x, y)$ is a bivariate probability density function, the associated *joint characteristic function* $M(u, v)$ is defined by

$$M_{XY}(u, v) = \int_{-\infty}^{\infty} \int_{-\infty}^{\infty} e^{iux} e^{ivy} f_{XY}(x, y)\, dx\, dy$$

a. Show that

$$i^{(n+m)}E\{X^nY^m\} = \left[\frac{\partial^{n+m}}{\partial u^n\,\partial v^m}M(u,v)\right]_{\substack{u=0\\v=0}}$$

b. Find $M(u, v)$ explicitly for the $f_{XY}(x, y)$ in Problems 1 and 2.

4. Let X_1, X_2, X_3, and X_4 be real random variables with a Gaussian joint probability. Show that

$$E\{X_1X_2X_3X_4\} = E\{X_1X_2\}E\{X_3X_4\}$$
$$+E\{X_1X_3\}E\{X_2X_4\} + E\{X_1X_4\}E\{X_2X_3\}$$

Hint: Use joint characteristic functions.

5. Let X be a Gaussian random variable with mean zero and unit variance. Let a new random variable Y be defined as follows: If $X = \xi$, then

$$Y = \begin{cases} \xi, & \text{with probability } \frac{1}{2} \\ -\xi, & \text{with probability } \frac{1}{2} \end{cases}$$

a. Determine the joint probability density of X and Y.

b. Determine the probability density of Y alone.

Note that although X and Y separately are Gaussian random variables, the joint probability density function for X and Y is not Gaussian.

Chapter 3

The Multidimensional Gaussian Distribution

The Joint Density Function

Let

$$\mathbf{C} = [c_{ij}], \qquad i = 1, 2, \ldots, n, \quad j = 1, 2, \ldots, n$$

be an $n \times n$ positive definite symmetric matrix. Let

$$\boldsymbol{\mu} = [\mu_1 \quad \mu_2 \quad \cdots \quad \mu_n]^T$$

be a vector with n real components. Let

$$\mathbf{X} = [X_1 \quad X_2 \quad \cdots \quad X_n]^T$$

be an n-dimensional random vector, that is, a vector having n components, each of which is a random variable. Let

$$\{a_1, a_2, \ldots, a_n\} \quad \text{and} \quad \{b_1, b_2, \ldots, b_n\}$$

be two sets of n real numbers each, with $a_i < b_i$ for each i, $1 \le i \le n$.

DEFINITION. The n-dimensional random vector $\mathbf{X}$ is said to possess an n-dimensional joint Gaussian distribution if it holds that

$$P\{a_i \le X_1 < b_1, a_2 \le X_2 < b_2, \ldots, a_n \le X_n < b_n\}$$
$$= \frac{|\mathbf{C}|^{-1/2}}{(2\pi)^{n/2}} \int_{a_1}^{b_1} \int_{a_2}^{b_2} \cdots \int_{a_n}^{b_n} \exp\{-\tfrac{1}{2}(\mathbf{x} - \boldsymbol{\mu})^T \mathbf{C}^{-1} (\mathbf{x} - \boldsymbol{\mu})\} \, |d\mathbf{x}| \quad (1)$$

Here $\mathbf{x} = [x_1 x_2 \quad \cdots \quad x_n]^T$ and $|d\mathbf{x}|$ denotes the differential volume element $dx_1 dx_2 \quad \cdots \quad dx_n$.

Theorem. Let $\mathbf{X}$ be an n-dimensional random vector with an n-dimensional joint Gaussian distribution. Then

$$E\{\mathbf{X}\} = \boldsymbol{\mu} \quad \text{and} \quad E\{(\mathbf{X} - \boldsymbol{\mu})(\mathbf{X} - \boldsymbol{\mu})^T\} = \mathbf{C}$$

where $\boldsymbol{\mu}$ and $\mathbf{C}$ are the parameter vector and matrix, respectively, in (1).

This is a well-known result, which we state without proof. A proof may be found in Ash (1970, Section 8.7).

We show in Appendix 1 that every covariance matrix $\mathbf{C}$ may be factored as

$$\mathbf{C} = \mathbf{L}\mathbf{D}\mathbf{L}^T \tag{2}$$

where $\mathbf{L}$ is lower triangular with all 1's on the main diagonal and $\mathbf{D}$ is diagonal. The inverse of this is

$$\mathbf{C}^{-1} = (\mathbf{L}^T)^{-1}\mathbf{D}^{-1}\mathbf{L}^{-1} \tag{3}$$

Now define

$$\mathbf{y} = \mathbf{L}^{-1}(\mathbf{x} - \boldsymbol{\mu}) \tag{4}$$

or, equivalently,

$$\mathbf{x} = \mathbf{L}\mathbf{y} + \boldsymbol{\mu} \tag{5}$$

Since $|\mathbf{L}| = 1$, we have $|d\mathbf{x}| = |d\mathbf{y}|$ and $|\mathbf{C}| = |\mathbf{D}|$. Therefore (1) becomes

$$P\{a_1 \le X_1 < b_1, a_2 \le X_2 < b_2, \ldots, a_n \le X_n < b_n\}$$

$$= \frac{|\mathbf{D}|^{-1/2}}{(2\pi)^{n/2}} \int_{\alpha_1}^{\beta_1} \int_{\alpha_2}^{\beta_2} \cdots \int_{\alpha_n}^{\beta_n} \exp[-\tfrac{1}{2}\mathbf{y}^T\mathbf{D}^{-1}\mathbf{y}]|d\mathbf{y}| \tag{6}$$

Thus, the integrand is reduced to the exponential of a weighted sum of squares. However, the limits $\{\alpha_i\}$ and $\{\beta_i\}$, $i = 1, 2, \ldots, n$, are not constants, so the integral does not simply reduce to a product of n one-dimensional integrals, but must be evaluated as an iterated integral. The exact determination of the $\{\alpha_i\}$ and $\{\beta_i\}$ and the details of the integration are explored in the problems.

Note that although we have reduced an arbitrary positive definite quadratic form to a sum of squares, the triangular factorization technique is different from the diagonalization of a symmetric matrix by means of finding eigenvalues and eigenvectors. That approach is an alternative method for evaluating (1). The triangular factorization is faster for numerical computation and has certain important features that we will examine in this book.

The Conditional Density Function

Let us turn our attention now to a completely different issue. Suppose we have a multidimensional joint Gaussian density function of a total of $n + m$ variables. For reasons having to do with certain applications, it is convenient to denote the first n variables by

$$\mathbf{x} = [x_1, x_2, \ldots, x_n]^T$$

and the remaining m variables by

$$\mathbf{y} = [y_1, y_2, \ldots, y_m]^T$$

We will denote the associated $(m + n) \times (m + n)$ covariance matrix by

$$\mathbf{C} = \begin{bmatrix} \mathbf{C}_{xx} & \mathbf{C}_{xy} \\ \mathbf{C}_{yx} & \mathbf{C}_{yy} \end{bmatrix} \tag{7}$$

where $\mathbf{C}_{xx}$ is $n \times n$, $\mathbf{C}_{xy}$ is $n \times m$, $\mathbf{C}_{yx} = \mathbf{C}_{xy}^T$, and $\mathbf{C}_{yy}$ is $m \times m$. The mean vector $\boldsymbol{\mu}$ is an $(m + n)$-dimensional vector:

$$\boldsymbol{\mu} = \begin{bmatrix} \boldsymbol{\mu}_x \\ \boldsymbol{\mu}_y \end{bmatrix} \tag{8}$$

In (8), $\boldsymbol{\mu}_x$ is n-dimensional and $\boldsymbol{\mu}_y$ is m-dimensional. With these agreements, our joint density function may be written

$$f(\mathbf{x}, \mathbf{y}) = \frac{|\mathbf{C}|^{-1/2}}{(2\pi)^{(n+m)/2}} \exp\left\{ -\tfrac{1}{2}\left[(\mathbf{x} - \boldsymbol{\mu}_x)^T (\mathbf{y} - \boldsymbol{\mu}_y)^T\right] \mathbf{C}^{-1} \begin{bmatrix} (\mathbf{x} - \boldsymbol{\mu}_x) \\ (\mathbf{y} - \boldsymbol{\mu}_y) \end{bmatrix} \right\} \tag{9}$$

If we perform marginal integration over $\mathbf{y}$ only, we will obtain the density function of $\mathbf{x}$ only:

$$f_1(\mathbf{x}) = \int_{-\infty}^{\infty} \cdots \int_{-\infty}^{\infty} f(\mathbf{x}, \mathbf{y}) |d\mathbf{y}| \tag{10}$$

Likewise, if we perform marginal integration over $\mathbf{x}$ only, we obtain the density function of $\mathbf{y}$ only:

$$f_2(\mathbf{y}) = \int_{-\infty}^{\infty} \cdots \int_{-\infty}^{\infty} f(\mathbf{x}, \mathbf{y}) |d\mathbf{x}| \tag{11}$$

The *conditional density function of x given y* is, by definition

$$f_c(\mathbf{x}|\mathbf{y}) = \frac{f(\mathbf{x},\mathbf{y})}{f_2(\mathbf{y})} \tag{12}$$

Since $f(\mathbf{x},\mathbf{y})$ and $f_2(\mathbf{y})$ are both Gaussian, taking their ratio merely means we subtract the quadratic form in the exponent of $f_2(\mathbf{y})$ from the quadratic form in the exponent of $f(\mathbf{x},\mathbf{y})$ and take the ratio of the numerical factors in front. Therefore $f_c(\mathbf{x}|\mathbf{y})$ will also be Gaussian. Even though it is a conditional density, it is nevertheless a probability density. That is, we may look at it as a density function of $\mathbf{x}$ alone, with $\mathbf{y}$ as a parameter.

Therefore we must have

$$\int_{-\infty}^{\infty} \cdots \int_{-\infty}^{\infty} f_c(\mathbf{x}|\mathbf{y})|d\mathbf{x}| = 1 \tag{13}$$

for all values of $\mathbf{y}$.

Hence $f_c(\mathbf{x}|\mathbf{y})$ must be of the form

$$f_c(\mathbf{x}|\mathbf{y}) = \frac{|\mathbf{P}|^{-1/2}}{(2\pi)^{n/2}} \exp\left[-\tfrac{1}{2}(\mathbf{x}-\mathbf{m})^T\mathbf{P}^{-1}(\mathbf{x}-\mathbf{m})\right] \tag{14}$$

Here $\mathbf{P}$ is an $n \times n$ matrix and $\mathbf{m}$ is an n-vector, both of which might possibly be functions of $\mathbf{y}$. It will turn out actually that $\mathbf{m}$ is a function of $\mathbf{y}$, but $\mathbf{P}$ is constant.

Our problem now is, given $f(\mathbf{x},\mathbf{y})$ as in (9), find $\mathbf{P}$ and $\mathbf{m}$ in terms of $\mathbf{C}_{xx}$, $\mathbf{C}_{xy}$, $\mathbf{C}_{yy}$, $\boldsymbol{\mu}_x$, $\boldsymbol{\mu}_y$, and $\mathbf{y}$. We begin by partitioning $\mathbf{C}^{-1}$ similarly to the way $\mathbf{C}$ is partitioned in (7). Let us write

$$\mathbf{C}^{-1} = \begin{bmatrix} \mathbf{A}_{xx} & \mathbf{A}_{xy} \\ \mathbf{A}_{yx} & \mathbf{A}_{yy} \end{bmatrix} \tag{15}$$

Matrix Inversion Lemma

Let us now carry out the following program:

1. Compute $\mathbf{A}_{xx}$, $\mathbf{A}_{xy}$, and $\mathbf{A}_{yy}$ in (15) explicitly in terms of $\mathbf{C}_{xx}$, $\mathbf{C}_{xy}$, and $\mathbf{C}_{yy}$.
2. Substitute the result of task 1 into (9).

3. Perform the marginal integration indicated in (11).
4. Form the ratio given for $f(\mathbf{x}|\mathbf{y})$ in (12) and equate it to the expression for $f_c(\mathbf{x}|\mathbf{y})$ given in (14).
5. Use the resulting equation to deduce explicit formulae for $\mathbf{P}$ in terms of $\mathbf{C}_{xx}$, $\mathbf{C}_{xy}$, and $\mathbf{C}_{yy}$, and for $\mathbf{m}$ in terms of these matrices and the vectors $\boldsymbol{\mu}_x$, $\boldsymbol{\mu}_y$, and $\mathbf{y}$.

Task 1 requires that we obtain the blocks in a partitioning of $\mathbf{C}^{-1}$ in terms of the blocks of a partitioning of $\mathbf{C}$. We will do this by simply applying the Gauss–Jordan method of matrix inversion to matrix blocks.

It will help to understand the procedure if we first do a very simple example to recall the basic procedure for 2×2 matrices. Suppose we want to invert the matrix

$$\begin{bmatrix} 5 & 3 \\ 3 & 2 \end{bmatrix}$$

We first form an augmented matrix by writing the 2×2 identity next to the given matrix, to obtain

$$\begin{bmatrix} 5 & 3 & 1 & 0 \\ 3 & 2 & 0 & 1 \end{bmatrix}$$

We now do a sequence of elementary row operations. First we reduce the block $\left[\begin{smallmatrix} 5 & 3 \\ 3 & 2 \end{smallmatrix}\right]$ to upper triangular form with 1's on the diagonal. We then continue elementary row operations to reduce this block to the 2×2 identity. When these operations are carried out on the augmented matrix, the desired matrix $\left[\begin{smallmatrix} 5 & 3 \\ 3 & 2 \end{smallmatrix}\right]^{-1}$ will appear on the right-hand side where $\left[\begin{smallmatrix} 1 & 0 \\ 0 & 1 \end{smallmatrix}\right]$ now is. Here is the sequence of operations, without further explanation:

$$\begin{bmatrix} 5 & 3 & 1 & 0 \\ 3 & 2 & 0 & 1 \end{bmatrix} \rightarrow \begin{bmatrix} 1 & \frac{3}{5} & \frac{1}{5} & 0 \\ 1 & \frac{2}{3} & 0 & \frac{1}{3} \end{bmatrix}$$

$$\rightarrow \begin{bmatrix} 1 & \frac{3}{5} & \frac{1}{5} & 0 \\ 0 & \frac{1}{15} & -\frac{1}{5} & \frac{1}{3} \end{bmatrix}$$

$$\rightarrow \begin{bmatrix} 1 & \frac{3}{5} & \frac{1}{5} & 0 \\ 0 & 1 & -3 & 5 \end{bmatrix}$$

$$\rightarrow \begin{bmatrix} 1 & 0 & 2 & -3 \\ 0 & 1 & -3 & 5 \end{bmatrix}$$

Hence

$$\begin{bmatrix} 5 & 3 \\ 3 & 2 \end{bmatrix}^{-1} = \begin{bmatrix} 2 & -3 \\ -3 & 5 \end{bmatrix}$$

It is important to realize that performing a sequence of row operations on a given matrix is equivalent to premultiplying the given matrix by a suitable transformation matrix. In the present procedure, the 2×2 block on the right-hand side is the pertinent transformation matrix at each step. For instance, to reduce the given matrix $\begin{bmatrix} 5 & 3 \\ 3 & 2 \end{bmatrix}$ to the triangular form $\begin{bmatrix} 1 & \frac{3}{5} \\ 0 & 1 \end{bmatrix}$, we should multiply $\begin{bmatrix} 5 & 3 \\ 3 & 2 \end{bmatrix}$ on the left by $\begin{bmatrix} \frac{1}{5} & 0 \\ -3 & 5 \end{bmatrix}$:

$$\begin{bmatrix} \frac{1}{5} & 0 \\ -3 & 5 \end{bmatrix} \begin{bmatrix} 5 & 3 \\ 3 & 2 \end{bmatrix} = \begin{bmatrix} 1 & \frac{3}{5} \\ 0 & 1 \end{bmatrix}$$

Finally, we may perform the inversion by an alternative sequence of row reductions whereby we first obtain a *lower* triangular matrix, as follows:

$$\begin{bmatrix} 5 & 3 & 1 & 0 \\ 3 & 2 & 0 & 1 \end{bmatrix} \to \begin{bmatrix} \frac{5}{3} & 1 & \frac{1}{3} & 0 \\ \frac{3}{2} & 1 & 0 & \frac{1}{2} \end{bmatrix}$$

$$\to \begin{bmatrix} \frac{1}{6} & 0 & \frac{1}{3} & -\frac{1}{2} \\ \frac{3}{2} & 1 & 0 & \frac{1}{2} \end{bmatrix}$$

$$\to \begin{bmatrix} 1 & 0 & 2 & -3 \\ \frac{3}{2} & 1 & 0 & \frac{1}{2} \end{bmatrix}$$

$$\to \begin{bmatrix} 1 & 0 & 2 & -3 \\ 0 & 1 & -3 & 5 \end{bmatrix}$$

The crucial point is now to realize that this procedure is still valid if our 2×2 matrix contains elements that are themselves block matrices of appropriate dimensions rather than scalars. Let us now apply this procedure with the objective of determining (15) from (7). We will do the lower triangular reduction first:

$$\begin{bmatrix} \mathbf{C}_{xx} & \mathbf{C}_{xy} & \mathbf{I}_n & \mathbf{0} \\ \mathbf{C}_{yx} & \mathbf{C}_{yy} & \mathbf{0} & \mathbf{I}_m \end{bmatrix}$$

$$\to \begin{bmatrix} \mathbf{C}_{xx} & \mathbf{C}_{xy} & \mathbf{I}_n & \mathbf{0} \\ \mathbf{C}_{yy}^{-1}\mathbf{C}_{yx} & \mathbf{I}_m & \mathbf{0} & \mathbf{C}_{yy}^{-1} \end{bmatrix}$$

$$\to \begin{bmatrix} \mathbf{C}_{xx} - \mathbf{C}_{xy}\mathbf{C}_{yy}^{-1}\mathbf{C}_{yx} & \mathbf{0} & \mathbf{I}_n & -\mathbf{C}_{xy}\mathbf{C}_{yy}^{-1} \\ \mathbf{C}_{yy}^{-1}\mathbf{C}_{yx} & \mathbf{I}_m & \mathbf{0} & \mathbf{C}_{yy}^{-1} \end{bmatrix}$$

Let us pause at this point to observe that we have derived the relationship

$$\begin{bmatrix} \mathbf{I}_n & -\mathbf{C}_{xy}\mathbf{C}_{yy}^{-1} \\ \mathbf{0} & \mathbf{C}_{yy}^{-1} \end{bmatrix}\begin{bmatrix} \mathbf{C}_{xx} & \mathbf{C}_{xy} \\ \mathbf{C}_{yx} & \mathbf{C}_{yy} \end{bmatrix} = \begin{bmatrix} \mathbf{C}_{xx} - \mathbf{C}_{xy}\mathbf{C}_{yy}^{-1}C_{yx} & \mathbf{0} \\ \mathbf{C}_{yy}^{-1}C_{yx} & \mathbf{I}_m \end{bmatrix} \tag{16}$$

Incidentally, in the above work, $\mathbf{I}_n$ denotes an $n \times n$ identity matrix and $\mathbf{I}_m$ denotes an $m \times m$ identity matrix.

For brevity, let us define

$$\mathbf{A}_{xx} = \left(\mathbf{C}_{xx} - \mathbf{C}_{xy}\mathbf{C}_{yy}^{-1}\mathbf{C}_{yx}\right)^{-1} \tag{17}$$

Our Gauss–Jordan reduction on blocks may now be continued as

$$\begin{bmatrix} \mathbf{A}_{xx}^{-1} & \mathbf{0} & \mathbf{I}_n & -\mathbf{C}_{xy}\mathbf{C}_{yy}^{-1} \\ \mathbf{C}_{yy}^{-1}\mathbf{C}_{yx} & \mathbf{I}_m & \mathbf{0} & \mathbf{C}_{yy}^{-1} \end{bmatrix}$$

$$\rightarrow \begin{bmatrix} \mathbf{I}_n & \mathbf{0} & \mathbf{A}_{xx} & -\mathbf{A}_{xx}\mathbf{C}_{xy}\mathbf{C}_{yy}^{-1} \\ \mathbf{C}_{yy}^{-1}\mathbf{C}_{yx} & \mathbf{I}_m & \mathbf{0} & \mathbf{C}_{yy}^{-1} \end{bmatrix}$$

$$\rightarrow \begin{bmatrix} \mathbf{I}_n & \mathbf{0} & \mathbf{A}_{xx} & -\mathbf{A}_{xx}\mathbf{C}_{xy}\mathbf{C}_{yy}^{-1} \\ \mathbf{0} & \mathbf{I}_m & -\mathbf{C}_{yy}^{-1}\mathbf{C}_{yx}\mathbf{A}_{xx} & \mathbf{A}_{yy} \end{bmatrix}$$

In the last block in the lower right-hand corner we have made the identification

$$\mathbf{A}_{yy} = \mathbf{C}_{yy}^{-1} + \mathbf{C}_{yy}^{-1}\mathbf{C}_{yx}\mathbf{A}_{xx}\mathbf{C}_{xy}\mathbf{C}_{yy}^{-1} \tag{18}$$

Therefore, we may first compute $\mathbf{A}_{xx}$ by means of (17), and then (18) enables us to find $\mathbf{A}_{yy}$. Moreover, we also have

$$\mathbf{A}_{xy} = -\mathbf{A}_{xx}\mathbf{C}_{xy}\mathbf{C}_{yy}^{-1} \tag{19}$$

and, of course, $\mathbf{A}_{yx} = \mathbf{A}_{xy}^T$.

Now let us carry out the procedure again, doing the upper triangular reduction first:

$$\begin{bmatrix} \mathbf{C}_{xx} & \mathbf{C}_{xy} & \mathbf{I}_n & \mathbf{0} \\ \mathbf{C}_{yx} & \mathbf{C}_{yy} & \mathbf{0} & \mathbf{I}_m \end{bmatrix}$$

$$\rightarrow \begin{bmatrix} \mathbf{I}_n & \mathbf{C}_{xx}^{-1}\mathbf{C}_{xy} & \mathbf{C}_{xx}^{-1} & \mathbf{0} \\ \mathbf{C}_{yx} & \mathbf{C}_{yy} & \mathbf{0} & \mathbf{I}_m \end{bmatrix}$$

$$\rightarrow \begin{bmatrix} \mathbf{I}_n & \mathbf{C}_{xx}^{-1}\mathbf{C}_{xy} & \mathbf{C}_{xx}^{-1} & \mathbf{0} \\ \mathbf{0} & \mathbf{C}_{yy} - \mathbf{C}_{yx}\mathbf{C}_{xx}^{-1}\mathbf{C}_{xy} & -\mathbf{C}_{yx}\mathbf{C}_{xx}^{-1} & \mathbf{I}_m \end{bmatrix}$$

Let us tentatively write

$$\mathbf{A}_{yy} = \left(\mathbf{C}_{yy} - \mathbf{C}_{yx}\mathbf{C}_{xx}^{-1}\mathbf{C}_{xy}\right)^{-1} \tag{20}$$

in analogy to (17), and we will momentarily disregard our previous result (18). Continuing then,

$$\begin{bmatrix} \mathbf{I}_n & \mathbf{C}_{xx}^{-1}\mathbf{C}_{xy} & \mathbf{C}_{xx}^{-1} & \mathbf{0} \\ \mathbf{0} & \mathbf{A}_{yy}^{-1} & -\mathbf{C}_{yx}\mathbf{C}_{xx}^{-1} & \mathbf{I}_m \end{bmatrix}$$

$$\rightarrow \begin{bmatrix} \mathbf{I}_n & \mathbf{C}_{xx}^{-1}\mathbf{C}_{xy} & \mathbf{C}_{xx}^{-1} & \mathbf{0} \\ \mathbf{0} & \mathbf{I}_m & -\mathbf{A}_{yy}\mathbf{C}_{yx}\mathbf{C}_{xx}^{-1} & \mathbf{A}_{yy} \end{bmatrix}$$

$$\rightarrow \begin{bmatrix} \mathbf{I}_n & \mathbf{0} & \mathbf{C}_{xx}^{-1} + \mathbf{C}_{xx}^{-1}\mathbf{C}_{xy}\mathbf{A}_{yy}\mathbf{C}_{yx}\mathbf{C}_{xx}^{-1} & -\mathbf{C}_{xx}^{-1}\mathbf{C}_{xy}\mathbf{A}_{yy} \\ \mathbf{0} & \mathbf{I}_m & -\mathbf{A}_{yy}\mathbf{C}_{yx}\mathbf{C}_{xx}^{-1} & \mathbf{A}_{yy} \end{bmatrix}$$

Therefore, since both methods of reduction must be valid, we have obtained the set of identities

$$\begin{aligned} \mathbf{A}_{xx} &= \left(\mathbf{C}_{xx} - \mathbf{C}_{xy}\mathbf{C}_{yy}^{-1}\mathbf{C}_{yx}\right)^{-1} \\ &= \mathbf{C}_{xx}^{-1} + \mathbf{C}_{xx}^{-1}\mathbf{C}_{xy}\mathbf{A}_{yy}\mathbf{C}_{yx}\mathbf{C}_{xx}^{-1} \end{aligned} \tag{21}$$

$$\mathbf{A}_{xy} = -\mathbf{A}_{xx}\mathbf{C}_{xy}\mathbf{C}_{yy}^{-1} = -\mathbf{C}_{xx}^{-1}\mathbf{C}_{xy}\mathbf{A}_{yy} \tag{22}$$

$$\mathbf{A}_{yx} = \mathbf{A}_{xy}^T \tag{23}$$

$$\begin{aligned} \mathbf{A}_{yy} &= \left(\mathbf{C}_{yy} - \mathbf{C}_{yx}\mathbf{C}_{xx}^{-1}\mathbf{C}_{xy}\right)^{-1} \\ &= \mathbf{C}_{yy}^{-1} + \mathbf{C}_{yy}^{-1}\mathbf{C}_{yx}\mathbf{A}_{xx}\mathbf{C}_{xy}\mathbf{C}_{yy}^{-1} \end{aligned} \tag{24}$$

These are the desired expressions for the blocks in (15). This result is known as the Matrix Inversion Lemma.

Conditional Mean and Covariance

We may now proceed with our objective of determining $\mathbf{P}$ and $\mathbf{m}$ in (14). Since $f_c(\mathbf{x}|\mathbf{y})$ is defined by (12), the next task is to compute $f_2(\mathbf{y})$. Although we could actually perform marginal integration of $f(\mathbf{x}, \mathbf{y})$ as indicated in

(11) to find $f_2(\mathbf{y})$, we may save ourselves the labor by thinking first. The joint density $f(\mathbf{x},\mathbf{y})$ must pertain to some $(n+m)$-dimensional random vector which, in consistency with our previous notation, must be denoted $[\mathbf{X}^T\mathbf{Y}^T]$. We then have the following facts:

$$E\left\{\begin{bmatrix}\mathbf{X}\\ \mathbf{Y}\end{bmatrix}\right\} = \int_{-\infty}^{\infty}\cdots\int_{-\infty}^{\infty}\begin{bmatrix}\mathbf{x}\\ \mathbf{y}\end{bmatrix} f(\mathbf{x},\mathbf{y})|d\mathbf{x}||d\mathbf{y}|$$

$$= \begin{bmatrix}\boldsymbol{\mu}_x\\ \boldsymbol{\mu}_y\end{bmatrix} \tag{25}$$

$$E\left\{\begin{bmatrix}\mathbf{X}-\boldsymbol{\mu}_x\\ \mathbf{Y}-\boldsymbol{\mu}_y\end{bmatrix}\left[(\mathbf{X}-\boldsymbol{\mu}_x)^T(\mathbf{Y}-\boldsymbol{\mu}_y)^T\right]\right\}$$

$$= \int_{-\infty}^{\infty}\cdots\int_{-\infty}^{\infty}\begin{bmatrix}\mathbf{x}-\boldsymbol{\mu}_x\\ \mathbf{y}-\boldsymbol{\mu}_y\end{bmatrix}\left[(\mathbf{x}-\boldsymbol{\mu}_x)^T(\mathbf{y}-\boldsymbol{\mu}_y)^T\right] f(\mathbf{x},\mathbf{y})|d\mathbf{x}||d\mathbf{y}|$$

$$= \mathbf{C} = \begin{bmatrix}\mathbf{C}_{xx} & \mathbf{C}_{xy}\\ \mathbf{C}_{yx} & \mathbf{C}_{yy}\end{bmatrix} \tag{26}$$

Thus in particular we have that

$$E\{\mathbf{Y}\} = \boldsymbol{\mu}_y \tag{27}$$

$$E\left\{(\mathbf{Y}-\boldsymbol{\mu}_y)(\mathbf{Y}-\boldsymbol{\mu}_y)^T\right\} = \mathbf{C}_{yy} \tag{28}$$

Now, since equation (1) and the theorem immediately following it must hold for every positive integer value of n, we see that necessarily

$$f_2(\mathbf{y}) = \frac{|\mathbf{C}_{yy}|^{-1/2}}{(2\pi)^{m/2}}\exp\left\{-\tfrac{1}{2}(\mathbf{y}-\boldsymbol{\mu}_y)^T\mathbf{C}_{yy}^{-1}(\mathbf{y}-\boldsymbol{\mu}_y)\right\} \tag{29}$$

Using the result along with (9) and (15) in (12) then yields

$$f_c(\mathbf{x}|\mathbf{y}) = \frac{|\mathbf{C}_{yy}|^{1/2}|\mathbf{C}|^{-1/2}}{(2\pi)^{n/2}}$$

$$\times\exp\left\{-\tfrac{1}{2}(\mathbf{x}-\boldsymbol{\mu}_x)^T\mathbf{A}_{xx}(\mathbf{x}-\boldsymbol{\mu}_x) - (\mathbf{x}-\boldsymbol{\mu}_x)^T\mathbf{A}_{xy}(\mathbf{y}-\boldsymbol{\mu}_y)\right.$$

$$\left.-\tfrac{1}{2}(\mathbf{y}-\boldsymbol{\mu}_y)^T\mathbf{A}_{yy}(\mathbf{y}-\boldsymbol{\mu}_y) + \tfrac{1}{2}(\mathbf{y}-\boldsymbol{\mu}_y)\mathbf{C}_{yy}^{-1}(\mathbf{y}-\boldsymbol{\mu}_y)\right\} \tag{30}$$

When the exponent in (30) is expanded and compared with the exponent in (14) by separately matching the terms that are quadratic in **x**, linear in **x**, and constant with respect to **x** (i.e., possibly functions of **y**), we obtain the identification

$$\mathbf{P}^{-1} = \mathbf{A}_{xx} \tag{31}$$

$$\mathbf{m} = \boldsymbol{\mu}_x + \mathbf{C}_{xy}\mathbf{C}_{yy}^{-1}(\mathbf{y} - \boldsymbol{\mu}_y) \tag{32}$$

Recall that $\mathbf{A}_{xx}$ was defined in (21). Also comparing (14) with (30) we see that we need to have the relation between determinants

$$|\mathbf{P}| = \frac{|\mathbf{C}|}{|\mathbf{C}_{yy}|}$$

Now recall the result (16). Utilize the fact that the determinant of a block triangular matrix is equal to the product of the determinants of the blocks on the diagonal. Hence, using (17),

$$|\mathbf{C}_{yy}^{-1}||\mathbf{C}| = |\mathbf{A}_{xx}^{-1}| \tag{33}$$

Since the reciprocal of a determinant is the determinant of the inverse matrix, (31) together with (33) yield the required result.

The conclusion, then, is that the conditional density defined in (12) is given by (14), with **P** and **m** given by (31) and (32).

Significance of the Conditional Mean: Introduction to Bayesian Estimation

For the purposes of this section, let us momentarily drop the Gaussian assumption on our probability distribution. Let **X** be an n-dimensional random vector, let **Y** be an m-dimensional random vector, and suppose **X** and **Y** have a joint probability distribution with a joint density function $f(\mathbf{x}, \mathbf{y})$.

We are still concerned with formulae (10)–(12), but we now are assuming that $f(\mathbf{x}, \mathbf{y})$ is arbitrary rather than specifically Gaussian.

In many physical situations involving random phenomena, not all of the relevant variables are directly accessible. We might model such a situation in the present context by letting **Y** represent the set of variables which we can actually observe or measure, while **X** represents the remaining variables which are required by our theory to describe the situation under study, but which we cannot directly measure. In such a case, it is often useful to compute an estimate of $\mathbf{X}$ from the observed $\mathbf{Y}$. Toward this end, introduce the notation

$$\hat{\mathbf{X}} = \mathbf{g}(\mathbf{Y}) \tag{34}$$

Here $\hat{\mathbf{X}}$ is a new random vector, our estimate of $\mathbf{X}$. $\hat{\mathbf{X}}$ is random because it is a function of another random vector, namely $\mathbf{Y}$. The function $\mathbf{g}$ represents our procedure for computing $\hat{\mathbf{X}}$ from $\mathbf{Y}$.

The question then arises, how do we decide on a useful estimation procedure, that is, what function $\mathbf{g}$ should we use? There is a whole subject called Estimation Theory which is a subset of the field of Statistics and is devoted to answering this question for a variety of circumstances. We will confine our attention to one specific answer here.

First of all, by making the assumption that the object we want to estimate, namely $\mathbf{X}$, is a random vector having a known joint distribution with the object we can observe, that is, $\mathbf{Y}$, we have already made an assumption that is actually rather drastic in view of the full range of a priori assumptions it is possible to make in statistical theory. The assumption we have made is called the *Bayesian* assumption in Statistics.

Having made that assumption, a popular way of selecting the function $\mathbf{g}$ is to require that the choice of $\mathbf{g}$ be *optimum* relative to some specified criterion. A frequently used way of specifying this criterion is to assign a numerical value to the *estimation error*, $\mathbf{X} - \hat{\mathbf{X}}$, that is, we introduce a function $L(\mathbf{X} - \hat{\mathbf{X}})$ called the loss function, typically chosen real-valued, non-negative, and convex, which represents the inconvenience or disutility of making an estimation error. In order to come out with a function $\mathbf{g}$ which is deterministically specified even though we are dealing with random quantities, we choose $\mathbf{g}$ to minimize the expected value of the loss function.

An analytically very convenient choice of loss function is the square of the Euclidean norm. This choice is called the *Minimum Mean Squared Error* (MMSE) criterion.

Naturally enough, if we invoke the Bayesian assumption and then adopt the MMSE criterion, the resulting estimator is called the optimum Bayesian MMSE estimator. We now proceed to derive it. From (34), we have

$$\begin{aligned} E\{L(\mathbf{X} - \hat{\mathbf{X}})\} &= E\{\|\mathbf{X} - \hat{\mathbf{X}}\|^2\} \\ &= E\left\{\|\mathbf{X} - \mathbf{g}(\mathbf{Y})\|^2\right\} \\ &= \int_{R^n}\int_{R^m} \|\mathbf{x} - \mathbf{g}(\mathbf{y})\|^2 f(\mathbf{x}, \mathbf{y})|d\mathbf{x}||d\mathbf{y}| \end{aligned} \tag{35}$$

Now expand (35) by using the definition of the norm, and (12):

$$E\{L(\mathbf{X}-\hat{\mathbf{X}})\}$$

$$= \int_{R^m}\left\{\int_{R^n}[\mathbf{x}^T\mathbf{x} - 2\mathbf{g}^T(\mathbf{y})\mathbf{x} + \mathbf{g}^T(\mathbf{y})\mathbf{g}(\mathbf{y})]\, f_c(\mathbf{x}|\mathbf{y})|d\mathbf{x}|\right\} f_2(\mathbf{y})|d\mathbf{y}| \quad (36)$$

Define $\mathbf{m}(\mathbf{y})$ and $V(\mathbf{y})$ as follows:

$$\mathbf{m}(\mathbf{y}) = \int_{R^n} \mathbf{x} f_c(\mathbf{x}|\mathbf{y})|d\mathbf{x}| \quad (37)$$

$$V(\mathbf{y}) = \int_{R^n} \|\mathbf{x} - \mathbf{m}(\mathbf{y})\|^2 f_c(\mathbf{x}|\mathbf{y})|d\mathbf{x}| \quad (38)$$

Use the identity

$$\mathbf{x}^T\mathbf{x} = [\mathbf{x} - \mathbf{m}(\mathbf{y})]^T[\mathbf{x} - \mathbf{m}(\mathbf{y})] + 2\mathbf{m}^T(\mathbf{y})[\mathbf{x} - \mathbf{m}(\mathbf{y})] + \mathbf{m}^T(\mathbf{y})\mathbf{m}(\mathbf{y})$$

to obtain

$$\int_{R^n} \mathbf{x}^T\mathbf{x} f_c(\mathbf{x}|\mathbf{y})|d\mathbf{x}| = V(\mathbf{y}) + \|\mathbf{m}(\mathbf{y})\|^2 \quad (39)$$

The other terms in (36) are easily evaluated using (37) and (13):

$$\int_{R^n} 2\mathbf{g}^T(y)\mathbf{x} f_c(\mathbf{x}|\mathbf{y})|d\mathbf{x}| = 2\mathbf{g}^T(\mathbf{y})\mathbf{m}(\mathbf{y}) \quad (40)$$

$$\int_{R^n} \|\mathbf{g}(\mathbf{y})\|^2 f_c(\mathbf{x}|\mathbf{y})|d\mathbf{x}| = \|\mathbf{g}(\mathbf{y})\|^2 \quad (41)$$

Therefore (36) becomes

$$E\{L(\mathbf{X}-\hat{\mathbf{X}})\} = \int_{R^m}[V(\mathbf{y}) + \|\mathbf{m}(\mathbf{y})\|^2 - 2\mathbf{g}^T(\mathbf{y})\mathbf{m}(\mathbf{y}) + \|\mathbf{g}(\mathbf{y})\|^2]\, f_2(\mathbf{y})|d\mathbf{y}|$$

$$= \int_{R^m} V(\mathbf{y}) f_2(\mathbf{y})|d\mathbf{y}| + \int_{R^m} \|\mathbf{m}(\mathbf{y}) - \mathbf{g}(\mathbf{y})\|^2 f_2(\mathbf{y})|d\mathbf{y}| \quad (42)$$

The first integral does not depend upon $\mathbf{g}(\mathbf{y})$. The second integral is evidently non-negative, and it can be made zero by choosing $\mathbf{g}(\mathbf{y}) = \mathbf{m}(\mathbf{y})$.

Therefore we have shown that the optimal Bayesian MMSE estimator is the function $\mathbf{m}(\mathbf{y})$ associated with the pertinent conditional density $f_c(\mathbf{x}|\mathbf{y})$, defined in (37), that is, the conditional mean.

We emphasize again that this result was derived for an *arbitrary*, not necessarily Gaussian, joint density $f(\mathbf{x}, \mathbf{y})$.

In the Gaussian case, the function $\mathbf{m}(\mathbf{y})$ is given by (32), as we have seen. Further, the function $V(\mathbf{y})$ defined in (38) does not depend on $\mathbf{y}$ in the Gaussian case. In fact, $V(\mathbf{y})$ is just the trace of the matrix $\mathbf{P}$ defined by (31) and (21). It is extremely significant for applications that $\mathbf{P}$ is constant, that is, not a function of $\mathbf{y}$. From (42), we see that the minimum value of the expected loss is

$$\int_{R^m} V(\mathbf{y}) f_2(\mathbf{y}) |d\mathbf{y}|$$

In the Gaussian case, therefore, the minimum mean squared error is given by the trace of $\mathbf{P}$.

In fact, $\mathbf{P}$ is not only the covariance of the conditional density $f_c(\mathbf{x}|\mathbf{y})$, but also the covariance of the error random variable $\mathbf{X} - \hat{\mathbf{X}}$. In general, these are two different objects, but in the Gaussian case they coincide.

Finally, this seems an appropriate point to discuss a subtlety regarding the conditional mean and, in general, the conditional expectation operator.

Suppose we observe the random vector $\mathbf{Y}$. As explained in Chapter 1, any random variable potentially may assume a range of possible values. The value it actually assumes in a particular instance depends on which sample point of the underlying sample space is the point pertaining to this instance. For some conceptual purposes it is useful to distinguish between the random variable, which is a mapping from the sample space to the range space, and the point of the range space corresponding to a particular realization. Thus, in a particular instance we might make the observation $\mathbf{Y} = \mathbf{y}$. The vector $\mathbf{y}$ is considered to be a deterministic quantity, some specific point in the range of $\mathbf{Y}$.

This distinction shows up in two different versions of the conditional expectation. If we wish to calculate the conditional expected value of $\mathbf{X}$ given that we have observed $\mathbf{Y} = \mathbf{y}$, we write $E\{\mathbf{X}|\mathbf{Y} = \mathbf{y}\}$. In this case, the resulting object is not random, but it is a function of the vector $\mathbf{y}$, which we might consider as a parameter of our random experiment. The function $\mathbf{m}(\mathbf{y})$ is as follows:

$$E\{\mathbf{X}|\mathbf{Y} = \mathbf{y}\} = \mathbf{m}(\mathbf{y}) \tag{43}$$

On the other hand, we may wish to discuss the random variable that is created by this operation without positing a specific outcome for our observation. In that case, we simply write $E\{\mathbf{X}|\mathbf{Y}\}$.

Therefore, $E\{\mathbf{X}|\mathbf{Y} = \mathbf{y}\}$ is not a random object, but $E\{\mathbf{X}|\mathbf{Y}\}$ *is*. Specifically, it will be a function of the random vector $\mathbf{Y}$, that is, $\mathbf{m}(\mathbf{Y})$. That is, $E\{\mathbf{X}|\mathbf{Y}\}$ is the same as the random vector that results from using the random vector $\mathbf{Y}$ as the argument of the function $\mathbf{m}$. This discussion is pertinent to our remarks surrounding (34), where we introduced the random vector $\hat{\mathbf{X}}$ as an estimate of $\mathbf{X}$.

Specifically, in the Gaussian case, we write

$$\mathbf{m} = \boldsymbol{\mu}_x + \mathbf{C}_{xy}\mathbf{C}_{yy}^{-1}(\mathbf{y} - \boldsymbol{\mu}_y)$$

which is the same as (32). If now we choose $\mathbf{g}(\mathbf{y}) = \mathbf{m}(\mathbf{y})$, then (34) becomes

$$\hat{\mathbf{X}} = \boldsymbol{\mu}_x + \mathbf{C}_{xy}\mathbf{C}_{yy}^{-1}(\mathbf{Y} - \boldsymbol{\mu}_y) \tag{44}$$

Problems

1. Use the algorithm from Appendix 1 to factor the matrix $\mathbf{C}$ as $\mathbf{C} = \mathbf{LDL}^T$, where

$$\mathbf{C} = \begin{bmatrix} 5 & 10 & 15 \\ 10 & 24 & 38 \\ 15 & 38 & 64 \end{bmatrix}$$

Use your results to find $\mathbf{C}^{-1}$. Check your work by direct matrix multiplication.

2. Repeat problem 1 for the case

$$\mathbf{C} = \begin{bmatrix} 8 & -4 & -4 & 6 \\ -4 & 4 & 2 & -4 \\ -4 & 2 & 4 & -6 \\ 6 & -4 & -6 & 10 \end{bmatrix}$$

3. Given a Gaussian random vector

$$\mathbf{Z}^T = [Z_1 \quad Z_2 \quad Z_3 \quad Z_4]$$

let its mean value be

$$\boldsymbol{\mu}_Z^T = [1 \quad 1 \quad 1 \quad 1]$$

Let the covariance matrix of **Z** be **C**, where

$$\mathbf{C} = \begin{bmatrix} 8 & -4 & -4 & 6 \\ -4 & 4 & 2 & -4 \\ -4 & 2 & 4 & -6 \\ 6 & -4 & -6 & 10 \end{bmatrix}$$

a. Let $\mathbf{X}^T = [Z_1 \quad Z_2]$ and $\mathbf{Y}^T = [Z_3 \quad Z_4]$. Find the conditional probability density for **X** given **Y**.

b. Find $\mathbf{C}^{-1}$.

4. Repeat problem 3 for the case

$$\mathbf{C} = \begin{bmatrix} 0.5 & 0.5 & 1.0 & 0.5 \\ 0.5 & 1.0 & 1.5 & 1.0 \\ 1.0 & 1.5 & 5.0 & 3.0 \\ 0.5 & 1.0 & 3.0 & 2.0 \end{bmatrix}$$

5. Given the jointly Gaussian random vectors **X** and **Y**, where **X** is of dimension n and **Y** is of dimension m, let their respective means be $\boldsymbol{\mu}_x$ and $\boldsymbol{\mu}_y$, and their joint covariance matrix be

$$\mathbf{C} = \begin{bmatrix} \mathbf{C}_{xx} & \mathbf{C}_{xy} \\ \mathbf{C}_{yx} & \mathbf{C}_{yy} \end{bmatrix}$$

Let $\hat{\mathbf{X}} = E\{\mathbf{X}|\mathbf{Y}\}$.

a. Compute the unconditional error covariance

$$E\{(\mathbf{X} - \hat{\mathbf{X}})(\mathbf{X} - \hat{\mathbf{X}})^T\}$$

b. Compute the conditional error covariance

$$E\{(\mathbf{X} - \hat{\mathbf{X}})(\mathbf{X} - \hat{\mathbf{X}})^T|\mathbf{Y} = \mathbf{y}\}$$

c). Show that the results of tasks a and b both equal

$$\mathbf{P} = \mathbf{C}_{xx} - \mathbf{C}_{xy}\mathbf{C}_{yy}^{-1}\mathbf{C}_{yx}$$

6. This problem is concerned with the details of performing the multiple integral in equation (6). Suppose you have a computer program that will compute the value of the one-dimensional integral

$$\frac{1}{\sqrt{2\pi d^2}} \int_\alpha^\beta e^{-y^2/2d^2}\, dy = J(\beta, d) - J(\alpha, d)$$

for any given numerical values of d, α and β, with $\beta \geq \alpha$. We are going to arrive at the conclusion that this resource is inadequate to evaluate (6).

Adopt the vector inequality notation $\mathbf{x} \geq \mathbf{a}$ to represent the simultaneous set of scalar inequalities $x_1 \geq a_1, x_2 \geq a_2, \ldots, x_n \geq a_n$. Consider the limits in (1), which we may now write $\mathbf{a} \leq \mathbf{x} \leq \mathbf{b}$. In making the change of variables (5), this becomes $\mathbf{a} \leq \mathbf{Ly} + \boldsymbol{\mu} \leq \mathbf{b}$, or

$$\mathbf{a} - \boldsymbol{\mu} \leq \mathbf{Ly} \leq \mathbf{b} - \boldsymbol{\mu} \qquad (*)$$

Since $\mathbf{L}$ is lower triangular with all 1's on the diagonal, let us denote its elements as follows:

$$\mathbf{L} = \begin{bmatrix} 1 & 0 & \cdots & \cdots & & 0 \\ l_{21} & 1 & 0 & \cdots & & 0 \\ l_{31} & l_{32} & 1 & 0 & \cdots & 0 \\ \cdots & \cdots & \cdots & \cdots & \cdots & \cdots \\ \cdots & \cdots & \cdots & \cdots & \cdots & \cdots \\ l_{n1} & l_{n2} & \cdots & \cdots & & 1 \end{bmatrix}$$

Now the vector inequality $(*)$ is the simultaneous set of scalar inequalities

$$\begin{gathered} a_1 - \mu_1 \leq y_1 \leq b_1 - \mu_1 \\ a_2 - \mu_n \leq y_2 + l_{21} y_1 \leq b_2 - \mu_2 \\ a_3 - \mu_3 \leq y_3 + l_{32} y_2 + l_{31} y_1 \leq b_3 - \mu_3 \\ \cdots\cdots\cdots\cdots\cdots\cdots\cdots\cdots\cdots\cdots \\ a_n - \mu_n \leq y_n + l_{n,n-1} y_{n-1} + \cdots + l_{n1} y_1 \leq b_n - \mu_n \end{gathered}$$

Show that the results of the integration in (6) with respect to y_n alone will be

$$\begin{gathered} J(b_n - \mu_n - l_{n,n-1} y_{n-1} - \cdots - l_{n1}, d_n) \\ -J(a_n - \mu_n - l_{n,n-1} y_{n-1} - \cdots - l_{n1} y_1, d_n) \end{gathered}$$

Of course, $y_1, y_2, \ldots, y_{n-1}$ are not numerical constants, but are variables. The integral with respect to y_{n-1} in (6) does not therefore have the simple integrand

$$\exp\left[-\frac{y_{n-1}^2}{2d_{n-1}}\right]$$

but rather has this multiplied by the function of y_{n-1} represented by J above.

Carry this problem as far as you wish to pursue it. You might specialize to $n = 2$ or $n = 3$ and give geometrical descriptions of the region of integration in **y** space represented by (*). You might suppose that the value of each variable y_k is quantized into discrete steps, so that the integral could be approximately evaluated by using the one-dimensional computer program repetitively. For example, if $n = 3$, let y_1 and y_2 each be quantized into 100 levels. The integrals with respect to y_1 and y_2 are now replaced by discrete sums over these levels. There will thus be 10^4 grid points (y_1, y_2) projected onto the (y_1, y_2) plane. The integral in (6) now becomes a weighted sum of 10^4 one-dimensional integrals.

Finally, suppose that the transformation matrix **L** in (5) was the modal matrix whose columns are the eigenvectors of **C**. In that case **L** is, in general, an arbitrary square matrix rather than lower triangular. Discuss what happens to the limits in this case when the integral (1) is converted into a form analogous to (6).

Chapter 4

Finite Random Sequences

Introduction

In this chapter we will investigate the properties of finite sequences of scalar Gaussian random variables. We have two purposes for leading students through such an investigation. The practical purpose is to develop the skills associated with performing calculations pertaining to such sequences. The theoretical purpose is to develop the ability of students to think conceptually about Gaussian stochastic processes. In order to facilitate the theoretical purpose, we are examining the situation of finite length sequences with the aim of achieving an intuitive understanding that can be generalized to infinite length sequences and then to continuous-time processes. This chapter has been written so that it can be used without first covering Chapter 3. For those who have covered Chapter 3, a few parts will be slightly repetitive.

In particular, on the level of theoretical understanding we will present two contrasting viewpoints: the *successive* point of view and the *simultaneous* point of view. If the student can become comfortable with both viewpoints in this section, it will be easier for him or her to think about the material that is presented in later sections, when we come to it.

The Successive Viewpoint

Consider a sequence of n Gaussian random variables. Denote the sequence by $X_1, X_2, \ldots, X_n$. The picture we have in mind for generating such a sequence is that a physical device, for example, a computer, outputs the members of the sequence successively at regular intervals, for example, once every millisecond.

First of all, we may examine a typical member of the sequence, X_k, in isolation from all the others. In that case we are simply looking at a single Gaussian r.v., familiar from Chapter 2:

$$P\{a \le X_k \le b\} = \frac{1}{\sqrt{2\pi\sigma_k^2}} \int_a^b \exp\left[\frac{(x-\mu_k)^2}{2\sigma_k^2}\right] dx \tag{1}$$

In equation (1), we are assuming that the mean and variance of X_k are, respectively, μ_k and σ_k^2. In order to provide a mathematical specification of our sequence, it is already clear that, at the very least, it will be necessary to supply the n means $\mu_1, \mu_2, \ldots, \mu_n$ and the n variances $\sigma_1^2, \sigma_2^2, \ldots, \sigma_n^2$. As we will soon see, in the most general case it is necessary to supply more parameters yet.

The class of sequences for which the set of means and variances alone does provide a complete specification is the class of sequences of mutually independent (m.i.) r.v.'s. For the sequence $X_1, X_2, \ldots, X_n$ to be mutually independent, the r.v.'s must be independent by pairs, by triples, by quadruples, and so on, according to the general definition of mutual independence. Because of the special properties of the Gaussian distribution, in that case pairwise independence automatically implies all of the higher-order conditions for independence. Moreover, as we have seen in the section on bivariate Gaussian distributions in Chapter 2, if two Gaussian r.v.'s are uncorrelated, they are automatically independent. Therefore, suppose that for all i and j, $1 \le i \le n$, $1 \le j \le n$, it holds that

$$E\{(X_i - \mu_i)(X_j - \mu_j)\} = 0 \qquad (i \ne j) \tag{2}$$

When (2) holds, we are dealing with a mutually independent Gaussian sequence.

Mutually independent sequences have good points and bad points. A good point is that a minimal number of parameters are required to specify the sequence. A bad point is that there is no way to do statistical inference on such a sequence. We are interested in developing the subject of statistical inference, which means the following: Given observations of only part of the sequence, make some inferences about the properties of the members that have not been observed. Clearly, if we measure X_2 and observe the particular numerical value $X_2 = 17$, this knowledge does not change what we are in a position to say about the unobserved r.v. X_3. If X_2 and X_3 are independent, before and after the measurement of X_2 all we can say is that X_3 has mean μ_3 and variance σ_3^2, unchanged.

Presently we will examine cases where measurement of X_2 will change our state of knowledge about X_3. The method for doing statistical inference in such cases will be one of the major focal points of interest in this course. First, however, we want to study the mutually independent case a little further, because one of the tools we are going to develop is a way of manufacturing correlated sequences from mutually independent ones.

Perhaps the most basic sequence is the one governed by equations (1) and (2), and for which, moreover, it holds that

$$\mu_k = 0 \quad \text{and} \quad \sigma_k^2 = 1, \qquad 1 \le k \le n \tag{3}$$

Such a sequence is called an *i.i.d.* (independent identically distributed) *sequence* and also (for reasons connected with the folklore of the subject) *unit white Gaussian noise.*

Because unit white Gaussian noise will play a special role in subsequent developments, let us adopt a special notation for it: $W_1, W_2, \ldots, W_k, \ldots, W_n$ denotes a sequence of independent Gaussian r.v.'s with parameters given by (3).

Let us now consider the general case when (2) does not hold, but rather, in addition to the n values of the means $\mu_1, \mu_2, \ldots, \mu_k, \ldots, \mu_n$, we are also given the $n(n+1)/2$ different values of the covariances:

$$C_{ij} = E\{(X_i - \mu_i)(X_j - \mu_j)\} \tag{4}$$

The reason there are $n(n+1)/2$ and not n^2 is because necessarily $C_{ij} = C_{ji}$. It is often convenient in practice to represent these covariance parameters as the elements of a matrix $\mathbf{C}$, called the covariance matrix:

$$\mathbf{C} = \begin{bmatrix} C_{11} & C_{12} & \cdots & C_{1n} \\ C_{21} & C_{22} & \cdots & C_{2n} \\ \vdots & \vdots & \vdots & \vdots \\ C_{n1} & C_{n2} & \cdots & C_{nn} \end{bmatrix} \tag{5}$$

Every legitimate covariance matrix is necessarily symmetric and positive definite. In Appendix 1 we present a proof that every such matrix can be factored as

$$\mathbf{C} = \mathbf{T}\mathbf{T}^T \tag{6}$$

where $\mathbf{T}$ is a lower triangular matrix and $\mathbf{T}^T$ is its transpose, and we present an algorithm for doing this. Written out, **T** looks like

$$\mathbf{T} = \begin{bmatrix} T_{11} & 0 & \cdots & 0 \\ T_{21} & T_{22} & \cdots & 0 \\ \vdots & \vdots & \vdots & \vdots \\ T_{n1} & T_{n2} & \cdots & T_{nn} \end{bmatrix} \tag{7}$$

We now demonstrate the relevance of unit white Gaussian noise. Suppose we have a computer program for generating samples of it, that is, we can get as many realizations of the sequence $W_1, W_2, \ldots, W_n$ as we like. Imagine that we ask the computer to print out 1000 such realizations. For concreteness, suppose that $n = 100$.

We would then have a printout of 100,000 numbers, arranged in the format of 1000 separate sequences, where each sequence contains 100 numbers. Let us index each of the sequences by the Greek letter ω, so now we will write $W_1(\omega), W_2(\omega), \ldots, W_{100}(\omega)$, where ω is an integer from 1 to 1000. Specifically, $W_1(1), W_2(1), \ldots, W_{100}(1)$ would be the members of the first sequence (of length 100) printed out, $W_1(17), W_2(17), \ldots, W_{100}(17)$ would be the members of the 17th sequence printed, and so on.

It is customary to refer to the entire collection of 1000 sequences as an *ensemble of realizations* of unit white Gaussian noise. Of course, all this terminology is just arbitrary bookkeeping and earmarking on our part. What the computer has actually done is simply to generate 100,000 successive values of a Gaussian r.v. with zero mean and unit variance, all mutually independent. We have chosen to format them into 1000 separate sequences. The reason for imagining this computer experiment is because we hope it will be an aid to conceptualizing how Gaussian processes work.

In Appendix 2 we give an actual computer program for carrying out a similar experiment, along with the results obtained from it and some discussion of their significance. If the reader cannot visualize the situation described here, then he or she may turn to Appendix 2 for a concrete illustration.

If we wanted to verify that our computer-generated random variables had the required properties, we would need a very large ensemble to compute expected values. An extremely important point that we wish to bring out here is that the expectation operator E always means averaging over the *ensemble* (i.e., averaging over ω in the present example as ω runs from 1 to 1000), *never* over a single sequence (i.e., averaging over k in the present example as k runs from 1 to 100 with ω held fixed).

Thus, to verify that

$$E\{W_2\} = 0, \qquad E\{W_2^2\} = 1 \tag{8}$$

we would experimentally compute

$$M_2 = \frac{1}{1000} \sum_{\omega=1}^{1000} W_2(\omega) \tag{9}$$

and compare it to 0, and compute

$$V_2 = \frac{1}{999} \sum_{\omega=1}^{1000} (W_2(\omega) - M_2)^2 \tag{10}$$

and compare it to 1. In Appendix 2 we discuss the statistics of this particular example in detail and explain why we use 999 and not 1000 in the denominator in equation (10).

Conversely, if we didn't know how the computer had been programmed, we could compute M_2 and V_2 as a practical method of "measuring" the mean and variance of the random variable W_2. We emphasize that averaging over an *ensemble* of many different realizations of the same random sequence is the only way to "measure" statistical properties such as mean and variance. Ensembles have statistical properties, whereas one individual realization of a random sequence does not.

If you imagine this ensemble of 1000 separate sequences, and then imagine that you go in blindfolded and pick out one sequence, what you then have is a concept of what the expression "random sequence" actually means. The whole sequence is to be thought of as one object, and a random sequence is a complete sequence picked at one draw from a collection of such sequences, just as in probability theory you may have considered drawing colored balls at random from an urn.

The need for having an ensemble in order to talk about statistical properties of our random sequence becomes even more evident if we want to verify *independence*. Suppose, for example, that we want to verify that the random variables W_2 and W_6 are independent. If we merely had one sequence (i.e., one realization) we would have one specific value of W_2 and one specific value of W_6. In that situation it is meaningless to talk about whether or not those two numbers are independent.

However, if we had our ensemble of 1000 realizations, and if we could believe in the Gaussian-ness of the distribution, it would suffice to verify that W_2 and W_6 are uncorrelated. For this purpose we could compute the sample covariance

$$C_{26}^{\text{SAMPLE}} = \frac{1}{999} \sum_{\omega=1}^{1000} [W_2(\omega) - M_2][W_6(\omega) - M_6] \tag{11}$$

Provided that C_{26}^{SAMPLE} was suitably near 0, we could reliably conclude that W_2 and W_6 are uncorrelated. In Appendix 2 we provide some discussion of what the words "suitably" and "reliably," as used here, actually mean.

At last we can now connect up the various ideas that have been presented so far. Suppose we had an ensemble of realizations of a unit white noise random sequence. Let us denote the entire ensemble as $\{W_k(\omega)\}$. For each fixed (ω), the index k runs from 1 to n. Suppose we want to generate an ensemble $\{X_k(\omega)\}$ of realizations of a Gaussian random sequence with different parameters. Specifically, it is required to have

$$E\{X_k\} = \mu_k$$

and

$$E\{(X_i - \mu_i)(X_j - \mu_j)\} = C_{ij} \tag{12}$$

Here is a recipe for making $\{X_k(\omega)\}$ from $\{W_k(\omega)\}$. Put the C_{ij} into a matrix, as in (5). Factor the matrix, as in (6), to obtain the parameters T_{ij} as in (7). Note that

$$T_{ij} = 0 \quad \text{for} \quad j > i \tag{13}$$

Now put

$$X_1 = T_{11}W_1 + \mu_1$$

$$X_2 = T_{21}W_1 + T_{22}W_2 + \mu_2$$

$$X_3 = T_{31}W_1 + T_{32}W_2 + T_{33}W_3 + \mu_3$$

and in general, put

$$X_k = \sum_{j=1}^{k} T_{ij}W_j + \mu_k \tag{14}$$

One of the homework problems is to verify that this recipe works.

An important point of this entire discussion is for students to think about this whole question: Given a Gaussian random sequence with known properties, use it to generate another Gaussian random sequence with specified properties. How are you going to verify experimentally that any procedure you come up with is actually working? Unless you start with an *ensemble* of realizations of the original random sequence, and use it to generate an *ensemble* of realizations of the specified random sequence, there is no way to "measure" experimentally the values of any statistical parameter.

The Simultaneous Viewpoint

We begin the discussion of the simultaneous viewpoint by again considering the simplest case: unit white Gaussian noise. Let the n random variables comprising the sequence be collected into an n-dimensional vector, denoted by $\mathbf{W}$:

$$\mathbf{W} = [W_1 \ W_2 \ \cdots \ W_n]^T \tag{15}$$

Consider the probability distribution of the r.v. W_1.

Since it is Gaussian with zero mean and unit variance, the pertinent probability density function is given by

$$\begin{aligned} P\{w_1 \le W_1 < w_1 + dw_1\} &= f(w_1)\, dw_1 \\ f(w_1) &= \frac{1}{\sqrt{2\pi}} e^{-w_1^2/2} \end{aligned} \tag{16}$$

Now consider the joint distribution of all of the r.v.'s $W_1, W_2, \ldots, W_n$. We are interested in the probability of the joint event.

$$\begin{aligned} &P\{w_1 \le W_1 < w_1 + dw_1;\ w_2 \le W_2 < w_2 + dw_2; \ldots;\ w_n \le w_n < W_n < w_n + dw_n\} \\ &\quad = P\{w_1 \le W_1 < w_1 + dw_1\} P\{w_2 < W_2 < w_2 + dw_2\} \\ &\quad\quad \cdots P\{w_n \le W_n < w_n + dw_n\} \end{aligned} \tag{17}$$

The right-hand side of equation (17), in which the probability of the joint event written on the left-hand side has been factored into a product of probabilities of individual events, is correct only because all of these r.v.'s are mutually independent. Since each of the individual events is given by the same density function as in (16), we have

$$\begin{aligned} &P\{w_1 \le W_1 < w_1 + dw_1;\ w_2 \le W_2 < w_2 + dw_2; \ldots;\ w_n \le W_n < w_n + dw_n\} \\ &\quad = f(w_1) f(w_2) \cdots f(w_n)\, dw_1\, dw_2 \cdots dw_n \\ &\quad = (2\pi)^{-n/2} e^{-(w_1^2 + w_2^2 + \cdots + w_n^2)/2}\, dw_1\, dw_2 \cdots dw_n \end{aligned} \tag{18}$$

Examination of equation (18) will reveal to the student why it has been found notationally and conceptually convenient to introduce the idea of a vector-valued random variable, that is, a random vector $\mathbf{W}$, as in (15), and the probability density function associated with it, which we denote by $f_{\mathbf{W}}(\mathbf{w})$. Explicitly,

$$f_{\mathbf{W}}(\mathbf{w}) = (2\pi)^{-n/2} e^{-\|\mathbf{w}\|^2/2} \tag{19}$$

Correspondingly we may introduce the vector mean

$$\boldsymbol{\mu}_{\mathbf{W}} = E\{\mathbf{W}\} = \int_{-\infty}^{\infty} \cdots \int_{-\infty}^{\infty} \mathbf{w} f_{\mathbf{W}}(\mathbf{w})\,|dw| \tag{20}$$

In (20), the symbol $|d\mathbf{w}|$ is a notational shorthand for the differential volume element $dw_1\, dw_2 \cdots dw_n$.

Moreover, we may introduce the covariance matrix

$$\begin{aligned} \mathbf{C}_{\mathbf{W}} &= E\left\{(\mathbf{W} - \boldsymbol{\mu}_{\mathbf{W}})(\mathbf{W} - \boldsymbol{\mu}_{\mathbf{W}})^T\right\} \\ &= \int_{-\infty}^{\infty} \cdots \int_{-\infty}^{\infty} (\mathbf{w} - \boldsymbol{\mu}_{\mathbf{W}})(\mathbf{w} - \boldsymbol{\mu}_{\mathbf{W}})^T f_{\mathbf{W}}(\mathbf{w})\,|d\mathbf{w}| \end{aligned} \tag{21}$$

Explicit evaluation yields, of course,

$$\boldsymbol{\mu}_{\mathbf{W}} = [0\,0\,\cdots\,0]^T \quad \text{(the zero vector)}$$

$$\mathbf{C}_{\mathbf{W}} = \begin{bmatrix} 1 & 0 & \cdot & \cdot & \cdot & 0 \\ 0 & 1 & \cdot & \cdot & \cdot & 0 \\ \vdots & \vdots & \vdots & \vdots & \vdots & \vdots \\ \cdot & \cdot & \cdot & \cdot & \cdot & \cdot \\ 0 & 0 & \cdot & \cdot & \cdot & 1 \end{bmatrix} \quad \text{(the identity matrix)} \tag{22}$$

Note that adopting the simultaneous viewpoint allowed us to introduce the concept of a probability distribution for the whole random sequence $W_1, W_2, \ldots, W_n$. This was done by looking at the sequence as a random vector. The desired distribution was then specified by providing the joint density function for all of the components of this random vector. Of course, this just returns us to the multidimensional Gaussian distribution and its properties, as covered in Chapter 3.

This ability to discuss the distribution of the whole sequence (that is, rather than discuss the sequence by considering one member at a time we consider the entire sequence itself as a random object and introduce a probability distribution for that object) is the advantage of the simultaneous viewpoint. Thus, the advantage is descriptive and mathematical.

On the other hand, the random sequence as we are going to experience it in real life *will* be a succession of scalar r.v.'s ordered in time. Hence, we need the successive viewpoint in order to discuss the situation as we actually experience it. Moreover, when the length of the sequences becomes infinite, the multidimensional distribution approach may become so unwieldy that the successive viewpoint is the only practical alternative.

The best strategy appears to be not to adopt either viewpoint exclusively, but rather to work with both and to develop the ability to shift gears and move from one to the other as appropriate. To analyze the theoretical properties of an ensemble of realizations of the random sequence, the simultaneous viewpoint is probably best. To look at the random sequence as an input to or output from a linear system, the successive viewpoint is probably more appropriate.

To gain still more familiarity with the simultaneous viewpoint let us again consider the situation discussed previously, of generating the random sequence (r.s.) $X_1, X_2 \ldots, X_n$ from the unit white noise sequence (u.w.n.s) $W_1, W_2, \ldots, W_n$.

Define the vectors

$$\mathbf{W} = \begin{bmatrix} W_1 \\ W_2 \\ \vdots \\ W_n \end{bmatrix}; \qquad \mathbf{X} = \begin{bmatrix} X_1 \\ X_2 \\ \vdots \\ X_n \end{bmatrix}; \qquad \boldsymbol{\mu} = \begin{bmatrix} \mu_1 \\ \mu_2 \\ \vdots \\ \mu_n \end{bmatrix}$$

and the matrix $\mathbf{C}$ as in equation (5). Then by definition of the u.w.n.s. we have

$$E\{\mathbf{W}\mathbf{W}^T\} = \mathbf{I} \tag{23}$$

By our problem statement (12), we are required to have

$$E\{\mathbf{X}\} = \boldsymbol{\mu}; \qquad E\{(\mathbf{X} - \boldsymbol{\mu})(\mathbf{X} - \boldsymbol{\mu})^T\} = \mathbf{C} \tag{24}$$

Let $\mathbf{C}$ be factorized as in (6), to obtain $\mathbf{T}$. Then our recipe for $\mathbf{X}$ is

$$\mathbf{X} = \mathbf{T}\mathbf{W} + \boldsymbol{\mu} \tag{25}$$

It is easy to verify that (24) follows from (6), (23), and (25). See problem 5 at the end of the chapter.

Lower Triangular Matrices and Causality

It is now appropriate to make a few remarks concerning the reason for specifying that the matrix $\mathbf{T}$ in (7) should be lower triangular. In index notation this condition appears as equation (13). In many applications, the index k will represent time, measured in some suitable discrete unit. The relationship (14) could be the input–output relationship of a linear system, where the sequence $\{W_j\}$ represents the stochastic portion of the input and the sequence $\{\mu_k\}$ represents the deterministic portion. The set of numbers T_{kj} in this context is sometimes called the "weighting pattern" of the system. Another term used for this is "influence coefficient." That is, T_{kj} represents the influence that an input occurring at time j will have on the output at time k. Since any actual physical system must be causal, that is, the past of the output may not depend on the future of the input, condition (13) is precisely the mathematical requirement that our system should be physically realizable.

The fact that the factorization (6) is always possible is a special case (for the situation of finite sequences of scalars) of a much more general result pertaining to stochastic processes. We will see some generalizations of this result in later chapters.

Problems

1. *An Integer-Valued Random Process*. Let the sample space Ω consist of 3^N points. Each point ω represents a sequence $(\omega_1, \omega_2, \dots, \omega_r, \dots, \omega_N)$ of independent, identically distributed random variables. Each random variable ω_r may assume one of the three values -1, 0, or $+1$, with respective probabilities $\frac{1}{4}, \frac{1}{2}, \frac{1}{4}$.

   The sample space Ω is thus the set of all sequences of length N which can be made from the three symbols -1, 0, $+1$. The above information specifies unambiguously a probability measure P on every subset of Ω. We are therefore now equipped with a probability trio $(\Omega, \mathscr{A}, P)$. Now define the random process as a sequence of random variables $\{X_1, X_2, \dots, X_N\}$ where

$$X_k = \sum_{j=1}^{k} \omega_j$$

   a. Find the probability distribution of the random variable X_4.

   b. Find the mean value function

$$m_k = E\{X_k\}, \qquad k = 1, 2, \dots, N$$

c. Find the covariance function

$$C_{jk} = E\{(X_j - m_j)(X_k - m_k)\} \quad j = 1,2,\ldots, N, \ k = 1,2,\ldots, N$$

2. This is based on the same process as in problem 1, but the questions are harder.

a. Find the probability of the event

$$\{X_1 > 0\} \cap \{X_2 > 0\} \cap \{X_3 > 0\}$$

b. Find a formula for the probability, as a function of k, of the event

$$\bigcap_{j=1}^{k} \{X_j > 0\}$$

c. Append an initial value $X_0 = 0$ to the sequence $\{X_k\}$. Define a random variable

$$T(\omega) = \min_{k>0} \{k\colon X_k = 0\}$$

$T(\omega)$ is the first time after zero that the X_k process equals zero again. See if you can think of a way to compute the probability distribution of $T(\omega)$.

3. Write a computer program using the program fragment given in Appendix 2 for a random number generator. The variable called RAND will be distributed uniformly on the interval [0, 1].

a. Devise a way to create 10 bins [0, .1), [.1, .2), ..., [.9, 1.0], and sort RAND into these bins.

b. Have the computer print out the number of values (out of 1000) of RAND that fell into each bin.

c. Calculate the sample mean and sample variance for 1000 samples of RAND. Compare with the theoretical values for a uniform distribution.

4. Let X_1 and X_2 be a pair of independent r.v.'s, each uniformly distributed on [0, 1]. Define

$$U_1 = \sqrt{-2\ln X_1}$$

$$U_2 = 2\pi X_2$$

$$G_1 = U_1 \cos U_2$$

$$G_2 = U_1 \sin U_2$$

Prove that G_1 and G_2 are independent r.v.'s, each being Gaussian with mean 0 and variance 1.

5. Let $\mathbf{C}$ be an $n \times n$ covariance matrix. Use the algorithm of Appendix 2 to factor it as $\mathbf{C} = \mathbf{LDL}^T$. Since $\mathbf{C}$ is positive definite, all of the elements of $\mathbf{D}$ are positive. Define $\mathbf{D}^{1/2}$ as the diagonal matrix obtained by replacing each element of $\mathbf{D}$ by its positive square root. Define $\mathbf{T} = \mathbf{LD}^{1/2}$. Then $\mathbf{C} = \mathbf{TT}^T$ as discussed in the text, where $\mathbf{T}$ has the appearance (7).

   Let μ_k, $k = 1, 2, \ldots, n$, be a set of real constants. Let W_k, $k = 1, 2, \ldots, n$, be a White Gaussian noise (WGN) sequence. Define a sequence of r.v.'s X_k, $k = 1, 2, \ldots, n$, by (14).

   Verify that the X_k so generated will have means and covariances given by (12).

   Suppose that the W_k are generated sequentially in time. Verify that (14) is in fact causal, that is, each X_k can be obtained as soon as the corresponding W_k is received. It is not necessary to wait until the entire WGN sequence has been received before it is possible to compute the X's. Show by constructing a specific example that this would not be the case if (13) did not hold.

Chapter 5

Stationary Random Sequences

Mean and Covariance Functions

Let us now suppose that we have a doubly infinite sequence of random variables X_k, where the index k is an integer ranging from $-\infty$ to $+\infty$. As in the previous chapter, we define the mean sequence μ_k by

$$\mu_k = E[X_k] \tag{1}$$

and the covariance elements c_{ij} by

$$c_{ij} = E\left[(X_i - \mu_i)(X_j - \mu_j)^T\right] \tag{2}$$

Also as in the previous chapter, we might visualize the μ_k as the components of a vector, and the c_{ij} as the elements of a matrix. However, the respective vector and matrix will now be infinite-dimensional.

Although the *concept* of this vector and this matrix remains just as important as in the previous chapter, there is much less facility in using the entire vector or matrix as an object in actual calculations. Indeed, most numerical calculations would have to be done element by element. Nevertheless, the ability to utilize both the simultaneous and the successive viewpoints, and to contrast the two perspectives, should be cultivated in the case of infinite sequences in order to maximize one's insight into any given situation.

From the simultaneous viewpoint, the entire sequence $\{X_k\}$ is just a single element (vector) in some suitably defined infinite-dimensional abstract space. Similarly, the covariance matrix **C** can be rigorously defined as an operator on a suitable infinite-dimensional abstract vector space.

A subclass of doubly infinite random sequences that is important, particularly in regard to applications, is the subclass of *stationary* random sequences. A heuristic interpretation of the concept of stationarity is that we cannot tell what time it is by looking at the statistical properties of some subset of the sequence $\{X_k\}$. For example, if we look at the subset corresponding to the index k in the range $-5000 \leq k \leq -3000$, the statistics of these 2001 r.v.'s will be no different from the statistics found for the case $+3000 \leq k \leq +5000$.

Now, the mean sequence $\{\mu_k\}$ defined in (1) is a deterministic sequence. If the value of μ_k in fact varies as k varies, then we could learn something about the value of k itself simply from knowing the value of μ_k and the law governing the variation. Therefore, to be stationary, a sequence $\{X_k\}$ must have the property that $E[X_k]$ is constant, independent of k. Another way of saying this is to require that

$$E[X_k] = E[X_{k+l}] \tag{3}$$

for all integer values of l and k. Equation (3) characterizes the property of being time invariant in terms of invariance with respect to time translation. That turns out to be the most useful way of characterizing stationarity: invariance under time translation of all statistical parameters of the random sequence.

Since a Gaussian sequence is completely specified if the mean sequence (1) and covariance elements (2) are all known, then if these are both invariant under time translation, the Gaussian sequence will be stationary. We have discussed the mean and written equation (3); the analogous equation for the covariance is

$$E[(X_i - \mu_i)(X_j - \mu_j)] = E[(X_{i+l} - \mu_{i+l})(X_{j+l} - \mu_{j+l})] \tag{4}$$

for all i, j, and l. If we assume (3) already holds, then (4) can be simplified to the condition

$$E[X_i X_j] = E[X_{i+l} X_{j+l}] \tag{5}$$

for all i, j, l.

If we now turn our attention toward the infinite-dimensional mean vector and covariance matrix, condition (3) means that all elements of the mean vector are the same, whereas condition (4) means that all of the elements of the covariance matrix which lie on a given diagonal parallel to the main diagonal must be equal.

For the analysis of stationary random sequences in situations where the index k represents time, it is useful to adopt a notation that can be applied to the continuous-time case as well as the discrete-time case. Indeed, there has been an implication all through the discussion so far that the index k represented time, although in fact this need not be the case. In what follows, however, we specifically assume that the index parameter of our random sequence is time, and we will now refer to the sequence as a *discrete-time random process*.

Given a stationary discrete-time random process $\{X(t)\}$, where t ranges over all of the integers, we define the *mean* μ_X by

$$\mu_X = E[X(t)] \tag{6}$$

and the *autocovariance* $c_{XX}(\tau)$ by

$$c_{XX}(\tau) = E\{[X(t+\tau) - \mu_X][X(t) - \mu_X]\} \tag{7}$$

Note well that in (7), the expectation operator E signifies an *ensemble* average, not a *time* average, as discussed in the previous chapter. It is because of the stationarity of the $\{X(t)\}$ process that the dependency on the variable t drops out, so that $c_{XX}(\tau)$ depends only on the shift parameter τ.

Example: White Noise Input to Discrete-Time System

As a first example, consider a unit WGN discrete-time random process $\{V(t)\}$, having zero mean. In this case, $\mu_V = 0$ and

$$c_{VV}(\tau) = \begin{cases} 1, & \tau = 0 \\ 0, & \tau \neq 0 \end{cases} \tag{8}$$

One way of generating a general stationary discrete-time random process from a WGN random process is by means of discrete-time convolution. We shall see presently just how general this method is. For now, it provides an example of the calculation of covariance functions. Let $h(t)$ be a bounded function of the discrete variable t, with the additional property that $\sum_{t=0}^{\infty}|h(t)|^2 < \infty$. It is only necessary for $h(t)$ to be defined on the non-negative integers. Consider the relation

$$X(t) = \sum_{n=0}^{\infty} h(n)V(t-n) \tag{9}$$

where $\{V\}$ is still unit WGN, with zero mean. Then

$$E[X(t)] = \sum_{n=0}^{\infty} h(n) E[V(t-n)] = 0 \tag{10}$$

In (10), the previous assumption of the square summability of h provides the basis for justifying the interchange of order of summation and expectation. Therefore the mean function for this $\{X(t)\}$ process is identically zero. In this case (7) simplifies to

$$c_{XX}(\tau) = E[X(t+\tau)X(t)] \tag{11}$$

Since the index of summation in (9) is a dummy variable, we may also write

$$X(t+\tau) = \sum_{l=0}^{\infty} h(l) V(t+\tau-l) \tag{12}$$

Substituting (12) and (9) into (11) yields

$$\begin{aligned} c_{XX}(\tau) &= E\left[\left(\sum_{l=0}^{\infty} h(l)V(t+\tau-l)\right)\left(\sum_{n=0}^{\infty} h(n)V(t-n)\right)\right] \\ &= E\left[\sum_{l=0}^{\infty}\sum_{n=0}^{\infty} h(l)h(n)V(t+\tau-l)V(t-n)\right] \\ &= \sum_{l=0}^{\infty}\sum_{n=0}^{\infty} h(l)h(n)E[V(t+\tau-l)V(t-n)] \end{aligned} \tag{13}$$

Now $(t+\tau-l)-(t-n) = (n+\tau-l)$, so

$$E[V(t+\tau-l)V(t-n)] = c_{VV}(n+\tau-l) \tag{14}$$

where c_{VV} is the function defined in (8). Therefore (13) becomes

$$c_{XX}(\tau) = \sum_{l=0}^{\infty}\sum_{n=0}^{\infty} h(l)h(n)c_{VV}(n+\tau-l) \tag{15}$$

If we perform the sum on l first, holding n fixed, we see that there is only one nonzero term, namely when $l = n + \tau$, and in this case $c_{VV} = 1$. Consequently we may substitute this value for l in $h(l)$, and remove c_{VV}

and the summation over l. Hence our final result is

$$c_{XX}(\tau) = \sum_{n=0}^{\infty} h(n+\tau)h(n) \tag{16}$$

It is clear from (11) that $c_{XX}(\tau)$ must be a symmetric function of τ, that is, $c_{XX}(-\tau) = c_{XX}(\tau)$, so that it is enough if (16) makes sense only for $\tau \geq 0$, since we may use symmetry to find $c_{XX}(\tau)$ for $\tau < 0$. However, rather than leaving $h(n)$ undefined for negative values of n, if we specifically assume that

$$h(n) = 0 \quad \text{whenever } n < 0 \tag{17}$$

then (16) is in fact consistent with the symmetry of c_{XX}. For instance, suppose specifically that $\tau = -5$. Then

$$c_{XX}(-5) = \sum_{n=0}^{\infty} h(n-5)h(n) \tag{18}$$

but because of (17), we may change the lower limit of the sum from 0 to 5. Thus

$$c_{XX}(-5) = \sum_{n=5}^{\infty} h(n-5)h(n)$$

Now make the change of index $m = n - 5$. When $n = 5$, $m = 0$, so we get

$$c_{XX}(-5) = \sum_{m=0}^{\infty} h(m)h(m+5) \tag{19}$$

On the other hand, substituting $\tau = 5$ into (16) yields directly

$$c_{XX}(5) = \sum_{n=0}^{\infty} h(n+5)h(n) \tag{20}$$

Clearly the right-hand sides of (19) and (20) are the same; hence

$$c_{XX}(-5) = c_{XX}(5) \tag{21}$$

Similarly, symmetry holds for every other value of τ.

This should serve as an example of the method of calculating an autocovariance function. We will now use the same example to illustrate another concept, the *cross-covariance function* of two jointly stationary random processes.

First of all, if $\{X(t)\}$ and $\{Y(t)\}$ are two general (not necessarily stationary) random processes, with respective means $\mu_X(t)$ and $\mu_Y(t)$, then their *cross-covariance* is defined as

$$c_{XY}(t_1, t_2) = E\{[X(t_1) - \mu_X(t_1)][Y(t_2) - \mu_Y(t_2)]\} \tag{22}$$

Even if $\{X(t)\}$ and $\{Y(t)\}$ individually are stationary, it can happen that they fail to be jointly stationary. For Gaussian processes, a necessary and sufficient condition is that $c_{XY}(t_1, t_2)$ should depend only on the difference $t_2 - t_1$. If this is the case, we introduce the shift parameter τ and write

$$c_{XY}(\tau) = E\{[X(t+\tau) - \mu_X(t+\tau)][Y(t) - \mu_Y(t)]\} \tag{23}$$

Returning now to our example introduced above in equation (9), let us calculate the cross-covariance $c_{XV}(\tau)$ between output and input. Since $\{X(t)\}$ and $\{V(t)\}$ both have zero mean,

$$c_{XV}(\tau) = E[X(t+\tau)V(t)] \tag{24}$$

From (9),

$$X(t+\tau) = \sum_{n=0}^{\infty} h(n)V(t+\tau-n) \tag{25}$$

Hence

$$\begin{aligned} c_{XV}(\tau) &= E\left[\sum_{n=0}^{\infty} h(n)V(t+\tau-n)V(t)\right] \\ &= \sum_{n=0}^{\infty} h(n)E[V(t+\tau-n)V(t)] \\ &= \sum_{n=0}^{\infty} h(n)c_{VV}(\tau-n) \end{aligned} \tag{26}$$

Therefore, $c_{XV}(\tau)$ is simply the output obtained when $c_{VV}(\tau)$ is applied as a deterministic input, that is, $c_{XV}(\tau)$ is the discrete-time convolution of $c_{VV}(\tau)$ with the function h.

We may proceed to recover our former result (16) by this method. We begin again with (11). Substitute (25) into (11) to obtain

$$\begin{aligned} c_{XX}(\tau) &= E\left[\sum_{n=0}^{\infty} h(n)V(t+\tau-n)X(t)\right] \\ &= \sum_{n=0}^{\infty} h(n)E[V(t+\tau-n)X(t)] \\ &= \sum_{n=0}^{\infty} h(n)c_{VX}(\tau-n) \end{aligned} \tag{27}$$

Now notice from (24) that

$$\begin{aligned} c_{VX}(\tau) &= E[V(t+\tau)X(t)] \\ &= E[X(t)V(t+\tau)] \\ &= E[X(t-\tau)V(t)] \\ &= c_{XV}(-\tau) \end{aligned} \tag{28}$$

where we have explicitly made use of the fact that $\{V\}$ and $\{X\}$ are both stationary processes. Therefore

$$c_{VX}(\tau-n) = c_{XV}(n-\tau) \tag{29}$$

From (26)

$$c_{XV}(n-\tau) = \sum_{\lambda=0}^{\infty} h(\lambda)c_{VV}(n-\tau-\lambda) \tag{30}$$

Using (29) and (30) in (27), we obtain

$$c_{XX}(\tau) = \sum_{n=0}^{\infty} h(n)c_{VX}(\tau-n) = \sum_{n=0}^{\infty}\sum_{\lambda=0}^{\infty} h(\lambda)h(n)c_{VV}(n-\tau-\lambda) \tag{31}$$

In (31), recalling (8), we see that the sum over n will contain only one nonzero term, namely when $n = \lambda + \tau$. Therefore (31) reduces to

$$c_{XX}(\tau) = \sum_{\lambda=0}^{\infty} h(\lambda)h(\lambda+\tau)$$

which is the same as (16).

Power Spectral Density and Bochner's Theorem

We now introduce a very important concept.

DEFINITION. Let $\{X(t)\}$ be a zero-mean stationary discrete-time random process with autocovariance function $c_{XX}(\tau)$. Its *power spectral density* $\phi_{XX}(\omega)$ is defined as

$$\phi_{XX}(\omega) = \sum_{\tau=-\infty}^{\infty} c_{XX}(\tau) e^{-i\omega\tau} \tag{32}$$

where omega ranges over the interval $-\pi \leq \omega \leq +\pi$.

Historically, this concept was originated, or at least popularized, by electrical engineers, who thought of $\{X(t)\}$ as a random electrical voltage waveform which might be transmitted over a piece of wire. Let this waveform be passed through a frequency filter, which only passes spectral components in a very narrow window of width $\Delta\omega$, centered at some frequency ω_0. In this case $\phi_{XX}(\omega_0)\,\Delta\omega$ represented, at least conceptually, the average power that would be dissipated if the output of the filter were fed into a 1-ohm resistor.

The question may arise, what if $\{X(t)\}$ has a nonzero mean? If $\{X(t)\}$ is to be stationary, then the mean must be *constant*, $\mu_X(t) = \mu_X$. In that case, to our electrical engineer, the mean μ_X simply represents a *d.c. component* of the random process, that is, we have actually a zero-mean random process connected in series with a deterministic voltage source of μ_X volts.

In practice, dealing with physical measurement of random processes, it is often more convenient to work with the *autocorrelation function* $r_{XX}(\tau)$, defined as

$$r_{XX}(\tau) = E[X(t+\tau)X(t)] \tag{33}$$

Here, (33) is taken as the definition even if the mean of $\{X\}$ is not zero. Writing

$$X(t) = [X(t) - \mu] + \mu$$

we see that

$$\begin{aligned} r_{XX}(\tau) &= E\{([X(t) - \mu] + \mu)([X(t+\tau) - \mu] + \mu)\} \\ &= E\{[X(t) - \mu][X(t+\tau) - \mu]\} + \mu^2 \\ &= c_{XX}(\tau) + \mu^2 \end{aligned} \tag{34}$$

In equation (32) we have defined the power spectral density as the discrete Fourier transform of the *autocovariance* function. In the literature the power spectral density is also defined as the discrete Fourier transform of the *autocorrelation function*. Let us examine what difference this makes. Adopt the notation

$$F[r_{XX}] = \sum_{\tau=-\infty}^{\infty} r_{XX}(\tau)e^{-i\omega\tau} \tag{35}$$

for the (discrete) Fourier transform of the autocovariance function.

Now the functions $\phi_{XX}(\omega)$ defined in (32) and $F[r_{XX}](\omega)$ defined in (35) are functions of the *continuous* variable ω, that is, the argument ω varies continuously over the interval $-\pi \leq \omega \leq +\pi$. What (32) actually represents, therefore, is the exponential Fourier series representation of the function $\phi_{XX}(\omega)$ on the interval $-\pi \leq \omega \leq +\pi$, and the values of $c_{XX}(\tau)$, as τ ranges over the integers, are in fact the values of the corresponding coefficients of the Fourier series. From the theory of Fourier series, therefore, we must have

$$c_{XX}(\tau) = \frac{1}{2\pi}\int_{-\pi}^{\pi} e^{i\omega\tau}\phi_{XX}(\omega)\,d\omega \tag{36}$$

Equation (36) is, in fact, the *inversion integral* for computing the autocovariance function that corresponds to a given power spectral density function.

Equation (34) shows that $F[r_{XX}]$ will differ from $\phi_{XX}(\omega)$ by the Fourier transform of the constant μ^2. The question is, what is the Fourier transform of a constant? Answer: it's a delta function.

Note that if we take the inverse transform of $\mu^2\delta(\omega)$, we get

$$\frac{1}{2\pi}\int_{-\pi}^{\pi} e^{i\omega\tau}\mu^2\delta(\omega)\,d\omega = \frac{\mu^2}{2\pi} \tag{37}$$

Equation (37) implies also that

$$\frac{1}{2\pi}\sum_{\tau=-\infty}^{\infty} e^{-i\omega\tau} = \delta(\omega) \tag{38}$$

We show directly that this is so. Let $f(\omega)$ be any continuous function defined on $-\pi \leq \omega \leq +\pi$. It then has a complex exponential Fourier series on that interval, which we may write as

$$f(\omega) = \sum_{n=-\infty}^{\infty} \alpha_n e^{+in\omega} \tag{39}$$

where the coefficients α_n are given by

$$\alpha_n = \frac{1}{2\pi}\int_{-\pi}^{\pi} f(\omega) e^{-in\omega}\, d\omega \tag{40}$$

We make use of the following *fact*: If it holds that

$$\int_{-\pi}^{\pi} f(\omega) g(\omega)\, d\omega = f(0) \tag{41}$$

for every continuous function $f(\omega)$, then $g(\omega) = \delta(\omega)$. See Bremmerman (1965), problem 11, p. 19.

Now consider, for any continuous $f(\omega)$,

$$\int_{-\pi}^{\pi} f(\omega)\left(\frac{1}{2\pi}\sum_{\tau=-\infty}^{\infty} e^{-i\omega\tau}\right) d\omega = \sum_{\tau=-\infty}^{\infty} \frac{1}{2\pi}\int_{-\pi}^{\pi} f(\omega)\, e^{-i\omega\tau}\, d\omega = \sum_{\tau=-\infty}^{\infty} \alpha_\tau \tag{42}$$

Substituting $\omega = 0$ into (39) yields

$$f(0) = \sum_{n=-\infty}^{\infty} \alpha_n \tag{43}$$

Comparing the right-hand sides of (42) and (43), we see that

$$\int_{-\pi}^{\pi} f(\omega)\left(\frac{1}{2\pi}\sum_{\tau=-\infty}^{\infty} e^{-i\omega\tau}\right) d\omega = f(0) \tag{44}$$

so that, by using the fact stated in connection with (41), we deduce (38).

From (32), (34), and (35) it now follows that

$$F[r_{XX}](\omega) = \phi_{XX}(\omega) + 2\pi\mu^2\delta(\omega) \tag{45}$$

Therefore, a reason for preferring to define $\phi_{XX}(\omega)$ as the Fourier transform of $c_{XX}(\tau)$ rather than of $r_{XX}(\tau)$ is that we avoid having to deal with a delta function in $\phi_{XX}(\omega)$ at $\omega = 0$ when the process $\{X(t)\}$ has a nonzero mean value.

Returning now to the heuristic concept of the power spectral density as the physical power dissipated in a resistor, we see that this implies that the function $\phi_{XX}(\omega)$ should never be negative:

$$\phi_{XX}(\omega) \geq 0, \qquad -\pi \leq \omega \leq +\pi \tag{46}$$

The question arises whether the function defined in (32) will actually have this property. There is a theorem due to Bochner which asserts that this is so. The autocovariance $c_{XX}(\tau)$ was defined in (7) for stationary processes. However, making the obvious analogy between (7) and (2), we see that we could reconstruct the infinite covariance matrix $\mathbf{C}$ from knowledge of the function $c_{XX}(\tau)$. We also know that $\mathbf{C}$ must be non-negative definite. The content of Bochner's theorem is that (46) is necessary and sufficient for the matrix $\mathbf{C}$ to be non-negative definite.

Digression: Review of Constant Parameter Discrete-Time Deterministic Linear System Theory

Let $y(t)$ be a *deterministic* discrete-time sequence where the time index t ranges over only the *non-negative integers*, $0 \leq t < +\infty$. Its *z-transform* $Y(z)$ is defined as the formal power series

$$Y(z) = \sum_{t=0}^{\infty} y(t) z^{-t} \tag{47}$$

The utility of this definition lies in its behavior when the sequence $y(t)$ is subjected to a time shift. To illustrate this, let us determine the resulting transform when the argument t is replaced by $t + 1$. Consider the following sequence of equations:

$$\begin{aligned}\sum_{t=0}^{\infty} y(t+1) z^{-t} &= \sum_{s=1}^{\infty} y(s) z^{-(s-1)} \\ &= z \sum_{s=1}^{\infty} y(s) z^{-s} = z[Y(z) - y(0)]\end{aligned} \tag{48}$$

This property makes the z-transform useful in analyzing sequences governed by linear difference equations. The simplest example is the first-order difference equation

$$y(t+1) + ay(t) = 0 \tag{49}$$

The problem is to find a sequence $y(t)$ that satisfies (49) for all non-negative integer values of t. It turns out that in order to make the solution unique, it is necessary to specify the initial value:

$$y(0) = c \tag{50}$$

Applying (47) and (48) to (49) gives

$$z[Y(z) - y(0)] + aY(z) = 0 \tag{51}$$

Rearranging and using (50) yields

$$Y(z) = \frac{cz}{z + a} \tag{52}$$

Recall the Taylor series

$$\frac{1}{1 - x} = 1 + x + x^2 + x^3 + \cdots \tag{53}$$

which converges for $|x| < 1$. Proceeding strictly formally by analogy

$$\frac{z}{z + a} = \frac{1}{1 + \frac{a}{z}} = 1 - \frac{a}{z} + \frac{a^2}{z^2} - \frac{a^3}{z^3} + \cdots \tag{54}$$

this may be written

$$Y(z) = c \sum_{t=0}^{\infty} (-a)^t z^{-t} \tag{55}$$

Identifying this with (47) suggests that

$$y(t) = c(-a)^t \tag{56}$$

It is easy to verify directly that (56) is in fact the solution to (49) which satisfies (50).

Various results from the abstract theory of linear operators can be used to provide a rigorous basis for explaining why the preceding set of formal manipulations actually works, but we will not do that here. Our object is merely to give a cursory review in order to provide the concepts from the theory of z-transforms which we need for the analysis of stationary random sequences.

Suppose now we have a discrete-time linear system governed by a deterministic input–output relation analogous to (9):

$$y(t) = \sum_{n=0}^{\infty} h(n)u(t - n) \tag{57}$$

Here, $u(t)$, $h(t)$, and $y(t)$ are all deterministic discrete-time sequences defined on the non-negative integers. We introduce their respective transforms:

$$Y(z) = \sum_{t=0}^{\infty} y(t) z^{-t}$$

$$H(z) = \sum_{t=0}^{\infty} h(t) z^{-t}$$

$$U(z) = \sum_{t=0}^{\infty} u(t) z^{-t} \tag{58}$$

Let us now take the z-transform of (57):

$$\begin{aligned} Y(z) &= \sum_{t=0}^{\infty} z^{-t} \left(\sum_{n=0}^{\infty} h(n) u(t-n) \right) \\ &= \sum_{n=0}^{\infty} h(n) \sum_{t=0}^{\infty} u(t-n) z^{-t} \end{aligned} \tag{59}$$

Now let $r = t - n$. Then $t = r + n$. In order for (57) to make sense, we have to agree that the members of the sequence $u(\cdot)$ are all zero for negative values of the argument. In (59), when we change the index from t to r, we have to decide the appropriate range of summation using the new index r. Because of our agreement, we see that all of the nonzero terms in (59) will be included if we simply sum on r over the non-negative integers. Making this change, (59) becomes

$$\begin{aligned} Y(z) &= \sum_{n=0}^{\infty} h(n) \sum_{r=0}^{\infty} u(r) z^{-(r+n)} \\ &= \left(\sum_{n=0}^{\infty} h(n) z^{-n} \right) \left(\sum_{r=0}^{\infty} u(r) z^{-r} \right) \\ &= H(z) U(z) \end{aligned} \tag{60}$$

Therefore, input–output relation (57) in the time domain is converted into the simple product (60) by the z-transform.

The reason this result is significant for us is because (57) resembles the two equations (26) and (27) which we derived earlier. Having obtained this result, we end the *digression* and continue the study of random sequences.

Input–Output Relations for Spectral Densities

The next issue concerns comparing the definition of the power spectral density in (32) with the z-transform in (47). Both $c_{XX}(\tau)$ and $y(t)$ are deterministic discrete-time sequences. However, $y(t)$ is one-sided whereas $c_{XX}(\tau)$ is two-sided. The sum in (47) is singly infinite while the sum in (32) is doubly infinite. Finally, to get an expression resembling (32), the z in (47) would have to be replaced with $e^{i\omega}$.

In equation (23) we defined $c_{XY}(\tau)$, the cross-covariance of two jointly stationary processes. In analogy with (32), we now define their *cross-spectral density* $\phi_{XY}(\omega)$ as the discrete Fourier transform of c_{XY}:

$$\phi_{XY}(\omega) = \sum_{\tau=-\infty}^{\infty} c_{XY}(\tau)e^{-i\omega\tau} \tag{61}$$

Let us now derive the relationship between spectral densities which corresponds to (27). We have

$$\begin{aligned}\phi_{XX}(\omega) &= \sum_{\tau=-\infty}^{\infty} c_{XX}(\tau)e^{-i\omega\tau}\\ &= \sum_{\tau=-\infty}^{\infty} e^{-i\omega\tau}\left(\sum_{n=0}^{\infty} h(n)c_{VX}(\tau-n)\right)\\ &= \sum_{n=0}^{\infty} h(n) \sum_{\tau=-\infty}^{\infty} e^{-i\omega\tau}c_{VX}(\tau-n)\end{aligned} \tag{62}$$

Again, make the change of index $\lambda = \tau - n$. Since the range of summation for τ was doubly infinite, so will be the range for λ. Thus we have

$$\begin{aligned}\phi_{XX}(\omega) &= \sum_{n=0}^{\infty} h(n) \sum_{\lambda=-\infty}^{\infty} e^{-i\omega\lambda}e^{-i\omega n}c_{VX}(\lambda)\\ &= \left(\sum_{n=0}^{\infty} h(n)e^{-i\omega n}\right)\left(\sum_{\lambda=-\infty}^{\infty} c_{VX}(\lambda)e^{-i\omega\lambda}\right)\\ &= H(e^{i\omega})\phi_{VX}(\omega)\end{aligned} \tag{63}$$

In (63), $H(e^{i\omega})$ is the z-transform $H(z)$ defined in (58), with the variable z replaced by $e^{i\omega}$. The function $H(z)$, incidentally, is usually called the *sampled-data transfer function*.

An exactly parallel derivation shows that the spectral density relation corresponding to (26) is

$$\phi_{XV}(\omega) = H(e^{i\omega})\phi_{VV}(\omega) \tag{64}$$

Also, it is easy to show that the spectral density relation corresponding to (28) is

$$\phi_{VX}(\omega) = \phi_{XV}(-\omega) \tag{65}$$

Note that a *cross*-spectral density is not restricted to be non-negative; in fact, it may be complex-valued. Moreover, it possesses no particular symmetry property with respect to its argument.

A *power* spectral density, on the other hand, being the discrete Fourier transform of an *autocovariance*, is always non-negative as we mentioned earlier, hence always purely real, and moreover a symmetric function of its argument:

$$\phi_{XX}(\omega) = \phi_{XX}(-\omega) \tag{66}$$

Combining (63), (64), (65), and (66), we obtain the spectral density relation that corresponds to (31):

$$\phi_{XX}(\omega) = H(e^{i\omega})H(e^{-i\omega})\phi_{VV}(\omega) \tag{67}$$

The product $H(e^{i\omega})H(e^{-i\omega})$ is simply the squared magnitude of $H(e^{i\omega})$:

$$H(e^{i\omega})H(e^{-i\omega}) = |H(e^{i\omega})|^2 \tag{68}$$

Thus, (67) may be written

$$\phi_{XX}(\omega) = |H(e^{i\omega})|^2\phi_{VV}(\omega) \tag{69}$$

In the particular case when $c_{VV}(\tau)$ is given by (8), it is obvious that

$$\phi_{VV}(\omega) = 1, \qquad -\pi \le \omega \le +\pi \tag{70}$$

In this case, (69) reduces to the discrete Fourier transform of (16), namely

$$\phi_{XX}(\omega) = |H(e^{i\omega})|^2 \tag{71}$$

In words, this means that when unit WGN is applied as the input to a time-invariant discrete-time linear system having sampled-data transfer function $H(z)$, the power spectral density of the output is given by (71). In order to find the autocovariance $c_{XX}(\tau)$ in this case, we could use (16), or it might be computationally easier to use (71) to find $\phi_{XX}(\omega)$ first, then use the inversion integral (36) to find $c_{XX}(\tau)$.

Factorization of Rational Spectral Densities

DEFINITION. A power spectral density function $\phi_{XX}(\omega)$ is called *rational* if and only if it can be written as a ratio of two polynomials in $e^{i\omega}$.

Since $\phi_{XX}(\omega)$ is necessarily real and even, it may also be expressed as a ratio of two polynomials in $\cos\omega$.

We have just shown that whenever we begin with a given linear system having transfer function $H(z)$ and apply WGN as its input, the power spectral density of the output is given by (71). Let us now inquire whether this proposition is reversible. It turns out that it is, provided ϕ_{XX} is rational. (It may happen to work sometimes when ϕ_{XX} is not rational, but we are not going to go into that.)

Theorem. Given a rational power spectral density function $\phi(\omega)$, there exists a rational function of z, $H(z)$, having poles inside the unit circle and zeros inside or on the unit circle in the complex z-plane such that $\phi(\omega) = |H(e^{i\omega})|^2$.

PROOF. The key steps of this proof consist of recognizing certain crucial algebraic facts. We content ourselves with pointing out what these facts are, and leave their detailed verification to the reader.

Fact 1. Since $\phi(\omega)$ is rational, real, and even, it can be written in the form

$$\phi(\omega) = \frac{\sum_{k=-m}^{m} \alpha_k e^{ik\omega}}{\sum_{l=-n}^{n} \beta_l e^{il\omega}} \tag{72}$$

where, moreover, $\alpha_k = \alpha_{-k}$ and $\beta_l = \beta_{-l}$, and all coefficients are real. Since the treatment of numerator and denominator is exactly the same, we henceforth consider only the numerator.

Fact 2. Since $\phi(\omega)$ is non-negative, the numerator can be factored as

$$A \prod_{k=1}^{m} (e^{i\omega} - z_k)(e^{-i\omega} - z_k^*) \tag{73}$$

where A is a real, positive constant, and the z_k are either real, or if complex, they occur in complex conjugate pairs. The asterisk denotes complex conjugate.

Fact 3. By choosing A properly, it is always possible to have all the z_k such that $|z_k| \leq 1$. (For the denominator, it must hold that the magnitude of each root is strictly less than 1.)

We give the demonstration of this fact by noting the identities

$$\begin{aligned}(e^{i\omega} - a) &= ae^{i\omega}\left(\frac{1}{a} - e^{-i\omega}\right) \\ (e^{-i\omega} - a) &= ae^{-i\omega}\left(\frac{1}{a} - e^{i\omega}\right)\end{aligned} \tag{74}$$

Combining all these facts, we conclude that $\phi(\omega)$ can always be written in the form

$$\phi(\omega) = K^2 \frac{\prod_{k=1}^{m}(e^{i\omega} - z_k)(e^{-i\omega} - z_k^*)}{\prod_{l=1}^{n}(e^{i\omega} - p_l)(e^{-i\omega} - p_l^*)} \tag{75}$$

where $|z_k| \leq 1$ for every k and $|p_l| < 1$ for every l. Now choose $H(z)$ as

$$H(z) = K\frac{\prod_{k=1}^{m}(z - z_k)}{\prod_{l=1}^{n}(z - p_l)} \tag{76}$$

and the proof is complete.

There is a system-theoretic result which is a direct consequence of this purely algebraic theorem, namely, given a stationary random sequence having rational power spectral density $\phi(\omega)$, it can be realized as the output of a time-invariant linear discrete-time dynamical system having discrete-time WGN as the input. If $\phi(\omega)$ is given, factor it as above and obtain $H(z)$. If WGN is applied as the input to a system having sampled-data transfer function $H(z)$, the output is a process with power spectral density $\phi(\omega)$, by the results of this chapter.

Equation (71) is the relation in the frequency domain which corresponds to equation (16) in the time domain. In (17), we made it explicit that $h(\cdot)$ is the impulse response of a *causal* linear system. We wish to conclude this chapter by pointing out the correspondence between the results obtained in this chapter and equation (6) of Chapter 4. In (6) of Chapter 4, we showed that a finite-dimensional covariance matrix can be factored as the product of a lower triangular matrix and its transpose. At the end of Chapter 4, we discussed the relationship between lower triangular matrices and causality. Thus, what we have done in this chapter is to extend the results of Chapter 4 to the infinite-dimensional case for stationary random sequences. In the next chapter, we extend these ideas to continuous-time processes.

Problems

1. Given the scalar homogeneous nonrandom difference equation $y(t+3) - 6y(t+2) + 11y(t+1) - 6y(t) = 0$ with initial conditions $y(1) = 6$, $y(2) = 10$, $y(3) = 24$, find the solution $y(t)$ by using z-transforms.
2. Let $\{V(\cdot)\}$ be unit white Gaussian noise, with zero mean. Consider the so-called *moving average* scheme with input V and output X:

$$X(t) = V(t) + bV(t-1)$$

   Determine (a) the autocovariance function of the output and (b) the cross-covariance of output and input.
3. Repeat problem 2 for the so-called *autoregressive* scheme

$$X(t) + aX(t-1) = V(t)$$

4. Repeat problem 3 for the complete autoregressive moving average (ARMA) scheme

$$X(t) + aX(t-1) = V(t) + bV(t-1)$$

   Try to solve it first by direct methods in the time domain. If you can't get it, then proceed by first finding the sampled data transfer function $H(z)$ for the system, use that to get the pertinent auto- and cross-spectral densities, and then use the inversion integral to get the desired auto- and cross-covariance functions.
5. A discrete-time linear system has the impulse response

$$h(t) = \left(\tfrac{1}{3}\right)^t + \left(\tfrac{1}{4}\right)^t, \qquad t \geq 0$$

   If the input is zero mean unit WGN, find the power spectral density and the autocovariance of the output.
6. A stationary discrete-time stochastic process has the power density function

$$\phi(\omega) = \frac{5 + 4\cos\omega}{10 + \cos\omega}$$

Perform the spectral factorization and determine the sampled-data transfer function of a stable system such that the output is the above process when the input is white noise.

7. Use Bochner's theorem to show that the function

$$c(\tau) = \begin{cases} \dfrac{\pi}{2}, & \tau = 0 \\ 1, & \tau = \pm 5 \\ 0, & \text{all other } \tau \end{cases}$$

cannot be the autocovariance of any random process.

8. Given a time-invariant discrete-time linear system with sampled-data transfer function

$$H(z) = \frac{z}{z^2 + 0.24z + 0.4}$$

If the input to this system is unit WGN with zero mean, find the power spectral density and autocovariance of the output.

9. Given the power spectral density function

$$\phi(\omega) = \frac{1 + \cos\omega}{2.125 - \cos 2\omega}$$

   a. Find the associated autocovariance function $c(\tau)$.
   b. If this represents the output from a time-invariant stable linear system driven by unit white noise, find the sampled data transfer function of the system.

10. The *same* WGN $\{V(\cdot)\}$ is simultaneously applied as input to two separate linear systems. The first has impulse response $g(t)$ and output $\{X(\cdot)\}$. The second has impulse response $h(t)$ and output $\{Y(\cdot)\}$. Find the cross-covariance $c_{XY}(\tau)$.

Chapter 6

Continuous-Time Stationary Gaussian and Second-Order Processes

Introduction

In this chapter we fulfill the development of a trend we have been following throughout this book. We began by examining scalar-valued random variables having a Gaussian probability distribution. We then studied finite-dimensional Gaussian random vectors, and in the previous chapter, Gaussian and second-order random infinite sequences. Now we allow the time index, which up until now has been discrete, to become continuous, and we study Gaussian and second-order random waveforms. We continue to use the convention of employing a capital letter to denote a random object, and so a typical random waveform will be denoted $X(t)$. In this book we will usually not distinguish notationally between a specific waveform and the entire family of waveforms from which it was drawn. Technically, it is the entire family of waveforms, sometimes called an ensemble of waveforms, together with its associated probability distribution, which is referred to by the term "stochastic process." In this book we will sometimes refer to "the stochastic process $X(t)$." If we really want to emphasize the point that we are dealing with an ensemble of waveforms and not just one single waveform, we may use curly brackets to indicate that we really have a set of waveforms, thus: $\{X(t)\}$. Indeed, since the use of the specific argument t suggests that we are looking at the value of the waveform at a particular time t and that what is inside the curly bracket is merely a plain old random variable, if we really want to emphasize that we have a set of functions, not a set of scalars, we may write, "the stochastic process $\{X(\cdot)\}$."

Even though, at first introduction, this object may seem somewhat unwieldy, if it is Gaussian, it is still completely characterized in terms of probability theory if we know its mean and covariance. Since time is now a continuous variable, the mean $\mu(t)$ is a real-valued function of the real variable t:

$$\mu(t) = E[X(t)] \tag{1}$$

The autocovariance $c(t, s)$ is a real-valued function defined on the plane:

$$c(t, s) = E\{[X(t) - \mu(t)][X(s) - \mu(s)]\} \tag{2}$$

Let $t_1, t_2, \ldots, t_n$ be any arbitrary set of n values of time. If we evaluated the stochastic process $\{X(\cdot)\}$ at those specific times and put the resulting values in an n-dimensional vector, we would have a random vector of the sort discussed previously, namely, $[X(t_1) \quad X(t_2) \quad \cdots \quad X(t_n)]$.

It should be clear that this vector, if it is Gaussian, will have a probability distribution exactly as written in equation (1) of Chapter 3, where the mean vector is given now by

$$\boldsymbol{\mu} = [\mu(t_1) \quad \mu(t_2) \quad \cdots \quad \mu(t_n)]^T \tag{3}$$

and the covariance matrix by

$$\mathbf{C} = \begin{bmatrix} c(t_1, t_1) & c(t_1, t_2) & \cdots & c(t_1, t_n) \\ c(t_2, t_1) & \cdots & \cdots & \cdots \\ \cdots & \cdots & \cdots & \cdots \\ c(t_n, t_1) & \cdots & \cdots & c(t_n, t_n) \end{bmatrix} \tag{4}$$

For each different selection of the value of n and for each choice of the values of the times $t_1, t_2, \ldots, t_n$, we will get a different vector $\boldsymbol{\mu}$ and a different matrix $\mathbf{C}$. All of them are associated with the one stochastic process $\{X(\cdot)\}$. In each case, the matrix $\mathbf{C}$ must be non-negative definite, in order to have a legitimate Gaussian distribution.

If all such matrices $\mathbf{C}$, resulting from all arbitrary choices of n and $t_1, t_2, \ldots, t_n$, are non-negative definite, then we say that the autocovariance $c(t, s)$ is a non-negative definite function. Obviously this is a very unwieldy property to test, as we have stated it. Further on in the chapter, for the special class of stationary processes, we will find a more convenient way of testing this property.

Since each individual waveform belonging to $\{X(\cdot)\}$ is a real-valued function of the real variable t, it may be investigated regarding the type of properties studied in real analysis, such as continuity and differentiability. For example, let t_0 be a fixed value of time. The process $\{X(\cdot)\}$ is called *mean-square continuous at* t_0 if and only if it holds that both

$$\lim_{t \to t_0} \mu(t) = \mu(t_0) \tag{5}$$

and

$$\lim_{t \to t_0} [c(t, t) - 2c(t, t_0) + c(t_0, t_0)] = 0 \tag{6}$$

Since all of the probabilistic analysis of Gaussian stochastic processes can be carried out knowing only the mean function $\mu(t)$ and the covariance function $c(t, s)$, a body of techniques has developed in practice whereby the only properties of these processes which are ever measured are the mean and the covariance, and the behavior of these processes in interacting with other processes, for example, passing through dynamic systems, is described in terms of means and covariances. This body of techniques can actually be applied to any stochastic process having suitable mean and covariance properties. In particular, it is not essential that the process be Gaussian.

Provided one does not need to make the sort of calculation of actual probabilities as is represented, for example, by the evaluation of the integral of equation (16) of Chapter 2 (that is, provided one is happy with the amount of information concerning the behavior of the process which is provided directly by the mean and the covariance alone, without invoking the underlying distribution), then indeed there is no reason to require that the process in question be Gaussian. Thus, a discipline and a body of subject matter exists and is referred to as "Second-Order Processes."

The techniques in this chapter can be applied equally well to true Gaussian processes or to the more general class of second-order processes. In working with this more general class, however, one must remember the caveat not to assume Gaussianness where it is not justified. For example, one should not attempt to calculate high-order moments of a random variable using known properties of the Gaussian distribution unless it is truly known that the random variable being studied actually is Gaussian. So long as one is working only with second-order processes in linear systems (the focus of most of this book), this issue probably will not arise. It does arise, however, if one attempts to extend the use of some of these techniques to nonlinear systems, as is done in Chapter 9.

Having made this warning and this disclaimer, the author feels he has met his responsibility in this matter, and it will not be mentioned again.

Covariance and Spectral Density Functions

Let us now focus our attention on the special class of stationary processes. A stochastic process is stationary only if *all* of its moments behave properly with regard to time shift. Since we will work only with the first two moments, if we are dealing with a second-order process not definitely known to be Gaussian, then what we will actually be defining here is the concept of "second-order stationarity."

DEFINITION. A second-order process is called stationary if and only if the following two conditions are met:

$$\mu(t) = \mu \tag{7}$$

$$c(t, s) = c(t - s) \tag{8}$$

In words, the mean is constant and the autocovariance depends only on the time difference.

Let us now introduce the notation designed specifically for stationary processes. Let $\{X(\cdot)\}$ be a zero-mean second-order stationary process. Let $\{Y(\cdot)\}$ be another such process. The autocovariance $c_{XX}(\tau)$ is defined by

$$c_{XX}(\tau) = E[X(t + \tau)X(t)] \tag{9}$$

Assuming that $\{X(\cdot)\}$ and $\{Y(\cdot)\}$ are jointly stationary, the cross-covariance $c_{XY}(\tau)$ is defined as

$$c_{XY}(\tau) = E[X(t + \tau)Y(t)] \tag{10}$$

Just as in the discrete-time case, it is also possible to introduce the autocorrelation functions $r_{XX}(\tau)$ and $r_{XY}(\tau)$. Recall the discussion associated with equations (33)–(45) of the previous chapter.

By definition of stationarity, $c_{XX}(\tau)$ does not depend on the value of t that is used in (9). To emphasize this, we could add some constant, say a, to t, and the result would be unchanged:

$$c_{XX}(\tau) = E[X(t + a + \tau)X(t + a)] \tag{11}$$

Also since we only have the product of two scalars under the expectation operator, the result does not depend on the order in which they are written:

$$c_{XX}(\tau) = E[X(t + a)X(t + a + \tau)] \tag{12}$$

Let us inquire what happens if τ is replaced by $-\tau$:

$$c_{XX}(-\tau) = E[X(t+a)X(t+a-\tau)] \tag{13}$$

This is easy to answer now. Since the value of a was arbitrary, equation (13) must still be true even if we set $a = \tau$:

$$c_{XX}(-\tau) = E[X(t+\tau)X(t)] \tag{14}$$

Comparing (14) with (9), we see that for an autocovariance,

$$c_{XX}(-\tau) = c_{XX}(\tau) \tag{15}$$

On the other hand, if we run through the same argument (try it yourself) with cross-covariance, we find

$$c_{XY}(-\tau) = c_{YX}(\tau) \tag{16}$$

Just as in the discrete-time case, analysis is greatly facilitated by the introduction of the concept of spectral density. In the continuous-time case, this is just the standard Fourier transform of the pertinent covariance function.

The power spectral density $\phi_{XX}(\omega)$ is given by

$$\phi_{XX}(\omega) = \int_{-\infty}^{\infty} c_{XX}(\tau) e^{-i\omega\tau}\, d\tau \tag{17}$$

From the theory of Fourier transforms, the inverse relation is given by

$$c_{XX}(\tau) = \frac{1}{2\pi}\int_{-\infty}^{\infty} \phi_{XX}(\omega) e^{i\omega\tau}\, d\omega \tag{18}$$

The cross-spectral density is given by

$$\phi_{XY}(\omega) = \int_{-\infty}^{\infty} c_{XY}(\tau) e^{-i\omega\tau}\, d\tau \tag{19}$$

and its inverse is

$$c_{XY}(\tau) = \frac{1}{2\pi}\int_{-\infty}^{\infty} \phi_{XY}(\omega) e^{i\omega\tau}\, d\omega \tag{20}$$

Let $\tau_1, \tau_2, \ldots, \tau_{n-1}$ be $n-1$ different values of the time shift parameter τ. Consider the $n \times n$ matrix

$$\mathbf{C} \equiv \begin{bmatrix} c_{XX}(0) & c_{XX}(\tau_1) & c_{XX}(\tau_2) & c_{XX}(\tau_{n-1}) \\ c_{XX}(\tau_1) & c_{XX}(0) & & \\ c_{XX}(\tau_{n-1}) & & & c_{XX}(0) \end{bmatrix} \tag{21}$$

This is just the covariance matrix $\mathbf{C}$ we already defined in (4), specialized to the stationary case. Physical realizability of the $\{X(\cdot)\}$ process requires that the matrix $\mathbf{C}$ be non-negative definite for every choice of n and every choice of $\tau_1, \tau_2, \ldots, \tau_{n-1}$. If this condition is met, the function $c_{XX}(\tau)$ itself is called non-negative definite.

The *Bochner theorem* states that $c_{XX}(\tau)$ is non-negative definite if and only if the power spectral density is non-negative at all frequencies:

$$\phi_{XX}(\omega) \geq 0, \qquad -\infty < \omega < +\infty \tag{22}$$

From basic properties of the Fourier transform, it follows from (15) that $\phi_{XX}(\omega)$ is necessarily real-valued. In fact, (15) actually implies

$$\phi_{XX}(-\omega) = \phi_{XX}(\omega) \tag{23}$$

Note that if we set $\tau = 0$ we obtain from (9) and (18)

$$E[X^2(t)] = \frac{1}{2\pi} \int_{-\infty}^{\infty} \phi_{XX}(\omega)\, d\omega \tag{24}$$

Therefore, unless the process $\{X(\cdot)\}$ is pathological, we must have

$$\int_{-\infty}^{\infty} \phi_{XX}(\omega)\, d\omega < \infty \tag{25}$$

Conversely, any function obeying (22), (23), and (25) is a possible power spectral density function.

Note that the cross-spectral density does not enjoy such remarkable properties. Since there is no particular relationship between $c_{XY}(\tau)$ and $c_{XY}(-\tau)$, the function $\phi_{XY}(\omega)$ need not be real, and it typically is complex valued. Needless to say, there is no particular symmetry relation between $\phi_{XY}(\omega)$ and $\phi_{XY}(-\omega)$ either.

From (19) and (16), what we can say is

$$\phi_{XY}(-\omega) = \int_{-\infty}^{\infty} c_{XY}(\tau) e^{i\omega\tau}\, d\tau = \int_{-\infty}^{\infty} c_{XY}(-\sigma) e^{-i\omega\sigma}\, d\sigma$$

$$= \int_{-\infty}^{\infty} c_{YX}(\sigma) e^{-i\omega\sigma}\, d\sigma = \phi_{YX}(\omega) \tag{26}$$

Laplace Transforms and Linear System Theory

Let $h(t)$ be the impulse response of a continuous-time time-invariant linear dynamic system. Its Laplace transform is defined by

$$H(s) = \int_0^{\infty} h(t) e^{-st}\, dt \tag{27}$$

$H(s)$ is called the *transfer function* of the system.

We need now to digress momentarily for the purpose of defining and clarifying the notational convention used in this book for Laplace and Fourier transforms. We will need to use the transfer function $H(s)$ from (27) with the argument s set equal to $i\omega$, which we will write $H(i\omega)$. Later in the book we will employ the Fourier transform on time functions defined for $-\infty < t < +\infty$. The defining integral will resemble (27), except the lower limit will be $-\infty$ rather than zero, and again the s in the exponent will be replaced by $i\omega$. There is no conflict in considering $H(i\omega)$ to be the *Fourier* transform of $h(t)$, because $h(t)$ is the impulse response of a causal system and consequently vanishes for all negative arguments. Therefore, this practice will cause no inconsistency.

We have also used the Fourier transform in (17), where we introduced the power spectral density $\phi_{XX}(\omega)$. In this case we have written the argument merely as ω, not $i\omega$. However, this is not inconsistent with the convention just stated, because we have used a *different letter*, namely ϕ, for the transform of the time function $c_{XX}(\tau)$. If we wanted to employ the above-given convention in this case, then we would have to use a *lowercase* letter for the time function, as is in fact the case (pardon the double entendre), and moreover denote the Fourier transform by the same letter capitalized, while using $i\omega$ as the argument, that is, $C_{XX}(i\omega)$. This latter usage is not standard in the literature, while the usage of the pair $c_{XX}(\tau) \leftrightarrow \phi_{XX}(\omega)$ is commonly found. These comments should clarify that this usage is consistent with usage of the pair $h(t) \leftrightarrow H(i\omega)$ (end of digression).

Let $u(t)$ and $y(t)$ be two deterministic functions, defined for $t \geq 0$. If they both are in the class of Laplace transformable functions, their respective Laplace transforms are given by

$$U(s) = \int_0^{\infty} u(t) e^{-st}\, dt \tag{28}$$

$$Y(s) = \int_0^{\infty} y(t) e^{-st}\, dt \tag{29}$$

Suppose that $u(t)$ is the input to a system with impulse response $h(t)$, and $y(t)$ is the corresponding output. In the time domain, the relationship among them is given by

$$y(t) = \int_0^{\infty} h(r)u(t-r)\,dr \tag{30}$$

In order to evaluate the integral in (30) for values of $r > t$, we adopt the convention that

$$u(t) = 0, \qquad -\infty < t < 0 \tag{31}$$

The relationship corresponding to (30) in the Laplace transform domain (s-domain) is, of course,

$$Y(s) = H(s)U(s) \tag{32}$$

Input–Output Relations for Stochastic Processes

Now suppose we have continuous-time second-order stationary process $\{X(\cdot)\}$ applied as the input to the system with impulse response $h(t)$, and let the resulting output process be $\{Y(\cdot)\}$. The relationship (30) holds, just as in the deterministic case:

$$Y(t) = \int_0^{\infty} h(r)X(t-r)\,dr \tag{33}$$

However, since $X(t) \neq 0$ for $-\infty < t < 0$, unlike (31), the Laplace transform cannot be applied directly to (33). As in the discrete-time case, let us first develop the corresponding input–output relations for the covariance functions, and then use the Fourier transform to obtain spectral density relationships.

We begin by computing $c_{XY}(\tau)$.

$$\begin{aligned}
c_{XY}(\tau) &= E[X(t+\tau)Y(t)] \\
&= E\left[X(t+\tau)\int_0^{\infty} h(r)X(t-r)\,dr\right] \\
&= \int_0^{\infty} h(r)E[X(t+\tau)X(t-r)]\,dr \\
&= \int_0^{\infty} h(r)c_{XX}(\tau+r)\,dr
\end{aligned} \tag{34}$$

Using (16) we obtain

$$c_{YX}(\tau) = \int_0^\infty h(r) c_{XX}(-\tau + r)\, dr \tag{35}$$

Finally, (15) allows to rewrite this as

$$c_{YX}(\tau) = \int_0^\infty h(r) c_{XX}(\tau - r)\, dr \tag{36}$$

The Fourier transform of (36) is

$$\phi_{YX}(\omega) = H(i\omega)\phi_{XX}(\omega) \tag{37}$$

From (26) and (23) we also have

$$\phi_{XY}(\omega) = H(-i\omega)\phi_{XX}(\omega) \tag{38}$$

For the autocovariance of the output, $c_{YY}(\tau)$, we obtain

$$\begin{aligned} c_{YY} &= E[Y(t+\tau)Y(t)] \\ &= E\left[Y(t+\tau)\int_0^\infty h(r)X(t-r)\, dr\right] \\ &= \int_0^\infty h(r)E[Y(t+\tau)X(t-r)]\, dr \\ &= \int_0^\infty h(r)c_{YX}(\tau + r)\, dr \end{aligned} \tag{39}$$

Since c_{YY} is an autocovariance it holds that $c_{YY}(\tau) = c_{YY}(-\tau)$, so we may write

$$c_{YY}(\tau) = \int_0^\infty h(r) c_{YX}(-\tau + r)\, dr \tag{40}$$

Again using (16), this becomes

$$c_{YY}(\tau) = \int_0^\infty h(r) c_{XY}(\tau - r)\, dr \tag{41}$$

The Fourier transform of (41) is

$$\phi_{YY}(\omega) = H(i\omega)\phi_{XY}(\omega) \tag{42}$$

Finally, combining (42) with (38) yields

$$\phi_{YY}(\omega) = H(i\omega)H(-i\omega)\phi_{XX}(\omega) \tag{43}$$

which we may also write as

$$\phi_{YY}(\omega) = |H(i\omega)|^2\phi_{XX}(\omega) \tag{44}$$

Spectral Factorization and the Paley–Wiener Criterion

Just as in the discrete-time case, we may develop a spectral factorization theorem and an associated realization procedure for a certain class of processes. In order to do that, we must first introduce continuous-time white noise. Now, discrete-time white noise was characterized by having the autocovariance given in equation (8) of Chapter 5, where τ was restricted to integer values. The associated power spectral density was given by

$$\phi_{VV}(\omega) = 1, \qquad -\pi \leq \omega \leq +\pi \tag{45}$$

The definition of white noise usually adopted in the continuous-time case is a process $\{V(\cdot)\}$ having zero mean and power spectral density

$$\phi_{VV}(\omega) = 1, \qquad -\infty < \omega < +\infty \tag{46}$$

in analogy with (45). Unfortunately, this has consequences which do not occur in the discrete-time case. Recalling the remark we made in connection with (25), we see that white noise is "pathological"

$$E[V^2(t)] = \int_{-\infty}^{\infty} \phi_{VV}(\omega)\, d\omega = \infty \tag{47}$$

Indeed, the reason for this result is apparent when we take the inverse Fourier transform of (46). The corresponding autocovariance function is

$$c_{VV}(\tau) = \delta(\tau) \tag{48}$$

the Dirac δ-function.

The point of view usually adopted is that although mathematical white noise as defined by (46) and (48) is not physically realizable, it is a very useful ideal concept for theoretical analysis. Moreover, it can be approximated closely enough by physical processes so as to be worthwhile experimentally.

Therefore, if the input process $\{X(\cdot)\}$ to our system were white noise, then the power spectral density of the output, by (44), is

$$\phi_{YY}(\omega) = |H(i\omega)|^2 \tag{49}$$

We know that every power spectral density must obey (22) and (23). The following question now arises: Does it follow from that alone, that it can always be factored like (49)?

In the discrete-time case, we answered this question only for the case of rational spectral densities. In the present case, we can give a more general answer. The answer is that, *no*, not every function obeying (22) and (23) alone can be factored like (49). However, every *rational* function can be factored, but the result is meaningless unless the spectral density obeys (25). If a rational function $\phi(\omega)$ obeys (22), (23), and (25), it can always be factored like (49), where $H(s)$ is the transfer function of a stable, causal linear system.

For an arbitrary nonrational function this is not enough. There is a criterion, called the *Paley–Wiener criterion*, which must be satisfied by a power spectral density function in order for it to be factored like (49), where $H(s)$ is the transfer function of a stable, causal linear system. Obviously if $\phi(\omega)$ is not rational, $H(s)$ will not be rational, so the resulting linear system would have an infinite-dimensional state space in that case.

The Paley–Wiener criterion is

$$\int_{-\infty}^{\infty} \frac{|\log \phi(\omega)|}{1+\omega^2}\, d\omega < \infty \tag{50}$$

The basic mathematical issue here belongs to the subject of analytic functions of a complex variable. The function $H(s)$ must be analytic on the $i\omega$ axis and in the entire right half of the complex s-plane, if it is the transfer function of a stable, causal system. For this reason, one cannot simply set

$$H(i\omega) = \sqrt{\phi(\omega)} \tag{51}$$

Although this does indeed fulfill (49) on the $i\omega$ axis, the resulting $H(i\omega)$ cannot be extended to a function $H(s)$ which is analytic on the right half-plane. In fact, (51) has to be replaced by the assumption

$$H(i\omega) = \sqrt{\phi(\omega)}\, e^{i\theta(\omega)} \tag{52}$$

where $\theta(\omega)$ is a suitably chosen phase angle function. The details of choosing $\theta(\omega)$ rest on the fact that the real and imaginary parts of $\log H(s)$ must be related by the Cauchy–Riemann conditions.

It turns out that it is not always possible to find a suitable $\theta(\omega)$ for a given $\phi(\omega)$. The criterion (50) is precisely the condition that $\phi(\omega)$ must satisfy in order for a suitable $\theta(\omega)$ to exist at all.

As mentioned above, whenever $\phi(\omega)$ is rational it is always possible to factor it so that $H(s)$ is analytic on the $i\omega$ axis and in the right half-plane. As a matter of fact, in order to get a unique factorization, we require not only that $H(s)$ should have no poles in the right half-plane but also that it should have no zeroes there. This is the same as saying that $\log H(s)$ should be analytic in the right half-plane.

In any event, the significance of this result for the stochastic realization problem is the same as in the discrete-time case. Whenever $\phi_{YY}(\omega)$ can be factored as in (49), with $\log H(s)$ analytic in the right half-plane, then the process $\{Y(\cdot)\}$ can be realized as the output of a linear system having transfer function $H(s)$ and driven by a white noise input.

As examples of functions which violate the Paley–Wiener criterion, let us now present the concepts of so-called band-limited white noise and narrow band noise. Band-limited white noise has a power spectral density that is constant over a (low-pass) frequency interval and is identically zero outside the interval:

$$\phi^{BL}(\omega) = \begin{cases} 1, & -\Omega \le \omega \le +\Omega \\ 0, & \text{elsewhere} \end{cases} \tag{53}$$

Narrow band noise will be recognized by communications engineers as the bandpass version of this spectrum:

$$\phi^{NB}(\omega) = \begin{cases} 1, & -\Omega \le |\omega - \omega_0| \le +\Omega \\ 1, & -\Omega \le |\omega + \omega_0| \le +\Omega \\ 0, & \text{elsewhere} \end{cases} \tag{54}$$

Let us find the autocovariance function $c^{BL}(\tau)$ corresponding to (53). From (18), we have

$$\begin{aligned} c^{BL}(\tau) &= \frac{1}{2\pi}\int_{-\infty}^{\infty} \phi^{BL}(\omega) e^{i\omega\tau}\, d\omega \\ &= \frac{1}{2\pi}\int_{-\Omega}^{\Omega} e^{i\omega\tau}\, d\omega = \frac{e^{i\Omega\tau} - e^{-i\Omega\tau}}{i\tau} \\ &= 2\Omega\left(\frac{\sin \Omega\tau}{\Omega\tau}\right) \end{aligned} \tag{55}$$

By the well-known properties of the Fourier transform of a modulated signal, we see that the autocovariance $c^{NB}(\tau)$ corresponding to (54) must be given by

$$c^{NB}(\tau) = 2c^{BL}(\tau)\cos \omega_0 \tau \tag{56}$$

Of course, we are assuming $\omega_0 \gg \Omega$.

Ergodic Processes

A concept that is very important for applications is that of an *ergodic* process. In order to introduce this concept, it will be helpful to recall the discussion in Chapter 4 of a computer simulation of discrete-time Gaussian white noise. There, we imagined a set of 100,000 random numbers, which we partitioned into 1000 strings of 100 numbers each. The time index was denoted by a subscript, from 1 to 100, while the ensemble variable was denoted by ω, which in this example was a discrete variable from 1 to 1000. The intent was to try to illustrate the meaning of the sample space (the 1000 separate strings, each string referenced by an individual value of ω) and to bring out that the expectation operator E always means to average over the ensemble of possible realizations.

This example illustrated what one might do in running a computer simulation. In a simulation one can, in principle, generate as many realizations as he wishes, that is, it is feasible to do a Monte Carlo type of experiment, and generate many versions of a sequence with the same statistical characteristics. However, when one is faced instead with processing data that have been obtained from the "real world," for example, from measurements taken on some actual physical process or device, there may be only *one* sequence. There may not be the luxury of repeating the same experiment many times in order to gather a large enough ensemble to permit ensemble averaging.

If statistical procedures are to be involved in such a situation with any degree of rigor, it is necessary to assume that the stochastic process under investigation is *stationary* and that a long time history of one waveform is available. The present chapter is devoted to the continuous-time case, so we will discuss waveforms rather than sequences henceforth.

The basic problem we now wish to consider, which might be faced by an engineer working in the area of applied stochastic processes, is the following. Given one long, but necessarily finite, data record of a continuous-time waveform assumed to be a typical representative of a particular stochastic process, determine the pertinent statistical parameters of the process. In the case of a Gaussian or second-order process, specifically, determine the mean and the autocovariance functions of the process. Let us consider the mean first.

Let the process be denoted $\{X(t)\}$. By definition,

$$\mu_X(t) = E\{X(t)\} \tag{57}$$

If the mean $\mu_X(t)$ really varies with time t, and if we have only one realization $X(\cdot)$ to work with, there is no way to estimate the mean, because we can't take an ensemble average over only one waveform. Clearly, the only way to have a set of numbers over which to average is to use the values of the waveform $X(\cdot)$ at different times. This says that the values of $X(\cdot)$ at different times, such as t_1, t_2, t_3, have some relation to the mean at yet another time, say $\mu_X(t_4)$.

There is only one assumption that workers in this field have found to have real utility here, namely, that the process $\{X(\cdot)\}$ is stationary and that the mean is some constant value, μ_X. If the data record extends over an interval $T_1 \le t \le T_2$, one possibility for estimating μ_X is to compute the time average

$$\bar{X}(T_1, T_2) = \frac{1}{T_2 - T_1} \int_{T_1}^{T_2} X(t)\, dt \tag{58}$$

It turns out that there is still no mathematical property of stationary Gaussian processes that allows one to conclude, in general, that $\bar{X}$ has any relation to μ_X. One specifically has to postulate that the process in question possesses an additional property that permits one to estimate or approximate μ_X by $\bar{X}$. That additional property is called *ergodicity*.

We will bypass the strict formal definition. Informally, a process first of all has to be stationary in order that ensemble averages be constant. If it is also ergodic, then it is legitimate to estimate parameters that are defined as ensemble averages by computing time averages.

Since a time average is often the only thing it is practical to compute in applications, the assumption of ergodicity is favored by those who have to come up with hard numbers based on real data. These workers also tend to prefer the use of the autocorrelation functions $r_{XX}(\tau)$ and $r_{XY}(\tau)$ mentioned earlier in this chapter [just before equation (11)]. For a process with zero mean, it makes no difference. However, at present we wish to consider a process with nonzero mean specifically to illustrate the meaning of ergodicity. Therefore, in contrast to (9), we define the autocorrelation function $r_{XX}(\tau)$ by

$$r_{XX}(\tau) = E[X(t+\tau)X(t)] \tag{59}$$

even when the mean of the process is not zero.

This function is still defined in terms of the ensemble average. Presently we will introduce a similar function defined in terms of the time average. First, however, we wish to resume our discussion of determining the mean μ_X by using a time averaging procedure.

The quantity denoted $\bar{X}(T_1, T_2)$ in (58) will, in general, depend upon which specific realization $X(\cdot)$, out of the ensemble of possible realizations $\{X(\cdot)\}$, was the one used under the integral to compute $\bar{X}(T_1, T_2)$. That means that $\bar{X}(T_1, T_2)$ is, itself, still a random object. Moreover, using different choices of interval, that is, different values of T_1, T_2, will in general yield different results.

In practice, there is no way to escape the fact that the object $\bar{X}(T_1, T_2)$ is going to possess these unstable features. Since $\bar{X}(T_1, T_2)$ is the only thing it is possible to compute, one computes it and adopts it as the best candidate for μ_X under the circumstances. The body of theory known as "ergodic theory" exists partly because theoreticians kindly attempted to supply practitioners with a suitable rationalization for this procedure.

One typically examines the probability distribution of $\bar{X}(T_1, T_2)$ and determines whether it possesses properties that permit the construction of this "suitable rationalization." If the process $\{X(\cdot)\}$ is truly Gaussian and not just second order, then since $\bar{X}(T_1, T_2)$ is obtained by a *linear* operation on $X(\cdot)$, the distribution of $\bar{X}(T_1, T_2)$ will be Gaussian also. Therefore, we completely know the distribution of $\bar{X}(T_1, T_2)$ if we know its mean and variance. Now, its mean is

$$E\{\bar{X}(T_1, T_2)\} = \frac{1}{T_2 - T_1}\int_{T_1}^{T_2} E\{X(t)\}\,dt = \frac{1}{T_2 - T_1}\int_{T_1}^{T_2} \mu_X\,dt = \mu_X \quad (60)$$

In (60), we have interchanged the order of the operations of expectation and time-integration. This can only be justified rigorously by using results from an advanced theory of integration called measure theory. We proceed with the assumption that the reader is not familiar with measure theory, and since it is not covered in this book, the reader will have to proceed on faith.

We see from (60) that the expected value of $\bar{X}(T_1, T_2)$ is, at least, the same as that of $X(t)$ for any t, namely μ_X. An object such as $\bar{X}(T_1, T_2)$, which is computed from the actual data and used as an estimate of some parameter of a probability distribution, is customarily called a *statistic*. When a statistic possesses the property represented by (60), namely, that its expected value is equal to the true value of the parameter, then the statistic is called *unbiased*.

Let us now compute the variance of $\bar{X}(T_1, T_2)$. Since

$$\begin{aligned}\operatorname{Var}(\bar{X}(T_1, T_2)) &= E\{[\bar{X}(T_1, T_2) - \mu_X]^2\} \\ &= E\{\bar{X}^2(T_1, T_2)\} - \mu_X^2 \end{aligned} \quad (61)$$

it is computationally easier to compute the second moment of $\bar{X}(T_1, T_2)$, and then subtract μ_X^2. Proceeding we have

$$E\{\bar{X}^2(T_1, T_2)\} = \frac{1}{(T_2 - T_1)^2} E\left\{\left(\int_{T_1}^{T_2} X(t)\,dt\right)^2\right\}$$

$$= \frac{1}{(T_2 - T_1)^2} E\left\{\int_{T_1}^{T_2}\int_{T_1}^{T_2} X(t)X(\lambda)\,dt\,d\lambda\right\}$$

$$= \frac{1}{(T_2 - T_1)^2}\int_{T_1}^{T_2}\int_{T_1}^{T_2} E\{X(t)X(\lambda)\}\,dt\,d\lambda$$

$$= \frac{1}{(T_2 - T_1)^2}\int_{T_1}^{T_2}\int_{T_1}^{T_2} r_{XX}(t - \lambda)\,dt\,d\lambda$$

$$= \frac{1}{(T_2 - T_1)^2}\int_{T_1}^{T_2}\int_{T_1}^{T_2} \left[c_{XX}(t - \lambda) + \mu_X^2\right] dt\,d\lambda$$

$$= \frac{1}{(T_2 - T_1)^2}\int_{T_1}^{T_2}\int_{T_1}^{T_2} c_{XX}(t - \lambda)\,dt\,d\lambda + \mu_X^2 \tag{62}$$

Combining (61) and (62),

$$\operatorname{Var}(\bar{X}(T_1, T_2)) = \frac{1}{(T_2 - T_1)^2}\int_{T_1}^{T_2}\int_{T_1}^{T_2} c_{XX}(t - \lambda)\,dt\,d\lambda \tag{63}$$

Since $c_{XX} \geq 0$ for all values of its argument, the integral in (63) will always be positive, and thus so will the variance of $\bar{X}(T_1, T_2)$.

If c_{XX} is actually known, and the integral can be evaluated, then we see that the hope of providing the desired rationalization for the use of $\bar{X}(T_1, T_2)$ rests on this variance being small. In the case of an unknown process, theoreticians shift their attention to a different statistic, specifically

$$\bar{X} = \lim_{T\to\infty} \frac{1}{2T}\int_{-T}^{T} X(t)\,dt \tag{64}$$

In general, it is *not* possible to obtain convergence results such as

$$\lim_{T_2\to\infty}\ \lim_{T_1\to-\infty} \bar{X}(T_1, T_2) = \mu_X \tag{65}$$

Unfortunately, in some cases the left-hand side of (65) may not converge at all. Thus, the practitioner who is stuck with using $\overline{X}(T_1, T_2)$ may not gain a great deal of reassurance even if $\overline{X}$ in (64) has favorable properties. If the integral in (63) grows slower than $(T_2 - T_1)^2$, as $T_2 \to +\infty$ and $T_1 \to -\infty$, then it is true that the variance of $\overline{X}(T_1, T_2)$ goes to zero, so $\overline{X}(T_1, T_2)$ converges to μ_X. However, as we said, if c_{XX} is unknown, this approach is unavailable.

When c_{XX} or r_{XX} has to be estimated as well as μ_X, it is customary to introduce the *time-average autocorrelation function* $R_{XX}(\tau)$, defined as

$$R_{XX}(\tau) = \lim_{T \to \infty} \frac{1}{2T} \int_{-T}^{T} X(t + \tau) X(t)\, dt \tag{66}$$

Our intention in this section was to discuss informally the concept of ergodicity. For further information, see Karlin and Taylor (1975). A rigorous definition for a stationary second-order process is that it is ergodic if and only if $\overline{X} = \mu_X$ and $R_{XX}(\tau) = r_{XX}(\tau)$. Since of course this is precisely the condition that cannot be determined in practice, this definition is of little use to the practitioner. What he does is boldly to *assume* the process he is dealing with is ergodic, and proceed from there, using "ergodic" as a buzzword when he writes up his results. We advise the reader to do the same.

Power Spectra for Deterministic Signals

In a number of areas of modern science and technology, the features of certain signals, that is, waveforms, are analyzed scrupulously with the intention of attempting to infer the properties of (a) some target from which the signal has been reflected or (b) some medium through which the signal has passed. Some examples are radar, sonar, meteorology, and geophysics. In some cases the signal is of natural origin, so the analyst must accept the underlying waveform as given and try to develop the most suitable signal processing technique for that particular category of waveform. In other cases, for example, active radar and sonar, the analyst has the luxury of engaging in a process of signal design for the purpose of constructing an optimum waveform for the task at hand.

In the case of naturally occurring signals, a common practice is to gather spectral data on these signals. One way of doing that is to calculate the time-average autocorrelation function defined in (66) and then take its Fourier transform. Other modern digital techniques, based on sampling the continuous-time waveform and then applying the FFT (Fast Fourier Transform) algorithm, are outside of the scope of this text. In any event, the objective is to characterize the waveforms as sample functions drawn from a second-order stationary stochastic process, with a particular shape of power spectrum, and then apply the methods of analysis just presented in this chapter.

In the case where the signal is devised by the system designer, it evidently is not a true stochastic signal, although some waveforms commonly used are deliberately chosen to resemble the waveforms of naturally occurring stochastic processes. Nevertheless, the analytical methods just presented have proven so useful that there has arisen the practice of applying the ideas of autocorrelation function and power spectral density to deterministic signals.

In order to work with a periodic signal and signals of finite duration, it proves convenient to distinguish between signals having finite *energy* and signals having finite *average power*. Given a deterministic signal $x(t)$ defined for $-\infty < t < +\infty$, we define its energy $\mathscr{E}$ by

$$\mathscr{E} = \int_{-\infty}^{\infty} x^2(t)\, dt \tag{67}$$

If $\mathscr{E} = \infty$, then we define the average power $\mathscr{P}$ by

$$\mathscr{P} = \lim_{T \to \infty} \frac{1}{2T} \int_{-T}^{T} x^2(t)\, dt \tag{68}$$

We need now to provide some definitions and discussion that will be used in Chapter 9. Let us restrict our attention to the case of finite energy deterministic signals. In analogy with the definition (66), we define the *autocorrelation function*

$$\rho_x(\tau) = \int_{-\infty}^{\infty} x(t)x(t+\tau)\, dt \tag{69}$$

Its Fourier transform, which we denote by $\phi_x(\omega)$, will be called the *energy spectrum of the waveform* $x(\cdot)$:

$$\phi_x(\omega) = \int_{-\infty}^{\infty} \rho_x(\tau) e^{-i\omega\tau}\, d\tau \tag{70}$$

Of course, the Fourier transform of the waveform $x(t)$ is conventionally designated as $X(i\omega)$:

$$X(i\omega) = \int_{-\infty}^{\infty} x(t) e^{-i\omega t}\, dt \tag{71}$$

Let us investigate the relationship between $X(i\omega)$ and $\phi_x(\omega)$. Substituting (69) into (70), we have

$$\phi_x(\omega) = \int_{-\infty}^{\infty}\int_{-\infty}^{\infty} x(t)x(t+\tau)e^{-i\omega\tau}\,dt\,d\tau \tag{72}$$

Let us perform the integration with respect to τ first. For this integral, substituting $\sigma = t + \tau$, we have

$$\int_{-\infty}^{\infty} x(t+\tau)e^{-i\omega\tau}\,d\tau = \int_{-\infty}^{\infty} x(\sigma)e^{i\omega t}e^{-i\omega\sigma}\,d\sigma$$

$$= e^{i\omega t}\int_{-\infty}^{\infty} x(\sigma)e^{-i\omega\sigma}\,d\sigma = e^{i\omega t}X(i\omega) \tag{73}$$

Substituting (73) into (72) yields

$$\phi_x(\omega) = \int_{-\infty}^{\infty} x(t)e^{i\omega t}X(i\omega)\,dt$$

$$= X(i\omega)\int_{-\infty}^{\infty} x(t)e^{i\omega t}\,dt = X(i\omega)X(-i\omega) = |X(i\omega)|^2 \tag{74}$$

So for finite energy deterministic signals, what we have called the energy spectrum turns out to be simply the squared magnitude of the Fourier transform of the signal itself.

We will complete this discussion by explaining the concept of the *matched filter*. Let a waveform $x(t)$ be given. Suppose it were possible to construct a filter having transfer function $X(-i\omega)$. If $x(t)$ were applied to this filter as an input, the Fourier transform of the output would obviously be $X(i\omega)X(-i\omega)$, that is, $\phi_x(\omega)$. Therefore the output as a function of time would be $\rho_x(t)$, the autocorrelation function of the waveform x.

Of course, this is physically unrealizable, as is evident from the symmetry property $\rho_x(t) = \rho_x(-t)$, which shows that the output is nonzero prior to $t = 0$. While we didn't place any restriction on $x(t)$, it is clear that if we do impose $x(t) = 0$ for $t < 0$, we still arrive at the same result. In that case the filter is producing output prior to the input, which shows that it is noncausal.

However, suppose that we make $x(t)$ be of finite duration T, that is, $x(t) = 0$ for $t < 0$ and for $t > T$. Now instead of requiring the filter to have transfer function $X(-i\omega)$, let us require it to have transfer function $X(-i\omega)e^{-i\omega T_1}$, where T_1 is a constant, $T_1 \geq T$. Under certain circumstances it is possible to find a waveform $x(t)$ such that $X(i\omega)$ and

$X(-i\omega)e^{-i\omega T_1}$ are both causal transfer functions. In that case, if $x(t)$ is the input to the filter, the Fourier transform of the output is $\phi_x(\omega)e^{-i\omega T_1}$ and the corresponding time function is $\rho_x(t - T_1)$. Since by our assumption on the finite duration of $x(\cdot)$ we have $\rho_x(\tau) = 0$ for $\tau > T$, this situation is physically realizable. A filter having the property that the output waveform is the autocorrelation of the input waveform except for a fixed delay is said to be *matched* to the input waveform. More details will be given in the Problems section.

The importance of this discussion for Chapter 9 is to introduce as a valid possibility the concept of having a waveform $x(t)$ which is of fairly constant, moderate amplitude over an interval $[0, T]$, but which possesses an autocorrelation function resembling a δ function. In that case, it is possible to transmit $x(t)$ and, by using a matched filter in the receiver, to obtain the same performance as if a δ function had been transmitted, but with much more moderate requirements for the peak power output of the transmitter.

Problems

1. Given a time-invariant continuous-time linear system, with impulse response $h(t) = e^{-t}\cos 2t$ for $t \geq 0$, let zero mean continuous-time WGN, with spectral density $= 1$, be applied as input. Denote the output process by $X(t)$. Find $c_{XX}(\tau)$ and $\phi_{XX}(\omega)$.
2. Let $\{X(\cdot)\}$ be a stationary zero mean continuous-time stochastic process. Define a new process $\{Z(\cdot)\}$ by

$$Z(t) = \int_{t-2}^{t} X(\sigma)\, d\sigma, \qquad -\infty < t < +\infty$$

   Calculate $c_{ZZ}(t, s) = E\{Z(t)Z(s)\}$. Is $\{Z(\cdot)\}$ a stationary process?
3. Let $\{X(\cdot)\}$ be a stationary zero mean continuous-time Gaussian process with power spectral density

$$\phi_{XX}(\omega) = \frac{|\omega|}{1 + \omega^4}$$

   Find the probability density function of the random variable $X(5)$.
4. The random variable θ is uniformly distributed on the interval $-\pi \leq \theta \leq +\pi$. Find the autocovariance $c_{XX}(\tau)$ and the power spectral density $\phi_{XX}(\omega)$ for the random process

$$X(t) = A\cos(\omega_0 t + \theta)$$

   where A and ω_0 are deterministic constants.

5. Let $\{X(\cdot)\}$ be a stationary zero mean process with power spectral density

$$\phi_{XX}(\omega) = \frac{25}{(\omega^2 + 4)(\omega^4 + 4)}$$

Find the transfer function $H(s)$ of a stable, causal, minimum phase linear system, such that when unit WGN is applied as input, the output process will be $\{X(\cdot)\}$.

6. Consider the power spectral density

$$\phi(\omega) = e^{-\omega^2/2}$$

Does it satisfy the Paley–Wiener criterion?

7. A linear system has an impulse response $h(t)$ of the form of a triangular pulse:

$$h(t) = \begin{cases} 0, & t < 0 \\ (1 - t), & 0 \le t \le 1 \\ 0, & t > 1 \end{cases}$$

A deterministic signal $x(t)$ is applied as input to this system. The signal $x(t)$ is given by

$$x(t) = \begin{cases} 0, & t < 0 \\ t, & 0 \le t \le 1 \\ 0, & t > 1 \end{cases}$$

Let the output of the system be denoted by $y(t)$.

a. Find $y(t)$.

b. Find the autocorrelation function

$$\rho_x(\tau) = \int_{-\infty}^{\infty} x(t)x(t + \tau)\, dt$$

c. Is this a matched filter?

d. Sketch $h(t)$, $x(t)$, $y(t)$, and $\rho_x(\tau)$.

Chapter 7

Nonstationary Continuous-Time Processes

Introduction

In this chapter we answer the question, to what extent can the ideas and concepts of the previous chapter be applied to nonstationary processes? We do not wish to keep the reader in suspense, so we will state the conclusions at once. The mean, auto- and cross-covariance, and impulse response functions can all be defined, if not exactly as in the previous chapter, then in a way which is a direct extension thereof. These functions remain the fundamental tools in the mathematical description and investigation of nonstationary second-order stochastic processes.

On the other hand, the frequency domain methods of analysis, that is, the Laplace and Fourier transforms and associated concepts such as transfer function and power spectral density function, have to be abandoned and replaced by direct time-domain techniques, specifically state-space methods. We will introduce these ideas by starting with some notions from the previous chapter, and showing first how the results for stationary processes can be recast in state-space form. It is then straightforward to generalize those results to nonstationary processes.

We begin with the input–output relation, equation (31) of Chapter 6, for deterministic time-invariant linear systems in Laplace transform format:

$$Y(s) = H(s)U(s) \tag{1}$$

Let us take the particular case when $H(s)$ is *rational*, which in the continuous-time situation means that $H(s)$ is a ratio of two polynomials in the variable s:

$$H(s) = \frac{b_0 s^m + b_1 s^{m-1} + \cdots + b_{m-1}s + b_m}{s^n + a_1 s^{n-1} + \cdots + a_{n-1}s + a_n} \tag{2}$$

From his or her knowledge of Laplace transform techniques, the reader will realize that this system can be described in the time domain by means of a linear differential equation with constant coefficients. Specifically, (1) and (2) imply that

$$\frac{d^n y(t)}{dt^n} + a_1 \frac{d^{n-1}y(t)}{dt^{n-1}} + \cdots + a_{n-1}\frac{dy(t)}{dt} + a_n y(t)$$

$$= b_0 \frac{d^m u(t)}{dt^m} + b_1 \frac{d^{m-1}u(t)}{dt^{m-1}} + \cdots + b_{m-1}\frac{du(t)}{dt} + b_m u(t) \tag{3}$$

For the time being, we will continue to confine the discussion to deterministic inputs and outputs. Let us review how to set up a state-space model for the system (3).

State Space Models

It is important to realize that state-space models are not unique. There are many different models for the same input–output relation. However, provided that the numerator and denominator of $H(s)$ in (2) contain no common factors, all state-space models of (3) which contain no superfluous features (so-called *minimal realizations*) will have the same dimension, namely n. See Balakrishnan (1983), Kailath (1980), Mortensen (1975), and Wiberg (1971) for a brief discussion of the concept of "state," as well as other references where further discussion can be found. It is assumed that the reader already has had a thorough introduction to state-space methods, so that a brief illustrative review will suffice to prepare him to extend his knowledge to systems with stochastic inputs and outputs.

We now describe one of the popular canonical forms for modeling (3). First of all we require an $n \times n$ matrix, denoted $\mathbf{A}$, having the negatives of the coefficients a_k from the left-hand side of (3) in its first column. The matrix $\mathbf{A}$ has all 1's on the diagonal above the main diagonal. Except for the

a_k's and 1's, every other element of **A** is 0. Its appearance is thus:

$$\mathbf{A} = \begin{bmatrix} -a_1 & 1 & 0 & \cdots & 0 \\ -a_2 & 0 & 1 & \cdots & 0 \\ \cdot & \cdot & \cdot & \cdots & \cdot \\ \cdot & \cdot & \cdot & \cdots & \cdot \\ \cdot & \cdot & \cdot & \cdots & \cdot \\ -a_{n-1} & 0 & 0 & \cdots & 1 \\ -a_n & 0 & 0 & \cdots & 0 \end{bmatrix} \tag{4}$$

Next we need an input vector, denoted **b**, of dimension n. The elements of **b** are the coefficients b_k on the right-hand side of (3), starting with b_m on the bottom and working up as far as the coefficients go. Since we explicitly assume $m < n$ here, it will always fit. Its appearance is thus:

$$\mathbf{b} = \begin{bmatrix} 0 \\ \cdot \\ \cdot \\ \cdot \\ b_0 \\ b_1 \\ \cdot \\ \cdot \\ \cdot \\ b_m \end{bmatrix} \tag{5}$$

The state vector $\mathbf{x}(t)$ is also a vector of dimension n, whose elements $x_1(t), x_2(t), \ldots, x_n(t)$ are functions of time.

$$\mathbf{x}(t) = \begin{bmatrix} x_1(t) \\ x_2(t) \\ \vdots \\ x_n(t) \end{bmatrix} \tag{6}$$

The output vector is denoted by **c**. For our canonical model, its leading element is 1 and all other elements are 0. We will use it in transposed form, thus:

$$\mathbf{c}^T = [1 \quad 0 \quad \cdots \quad 0] \tag{7}$$

With these definitions, our canonical state-space model of (3) is, of course,

$$\begin{aligned}\dot{\mathbf{x}}(t) &= \mathbf{A}\mathbf{x}(t) + \mathbf{b}u(t) \\ y(t) &= \mathbf{c}^T\mathbf{x}(t)\end{aligned} \tag{8}$$

By repeatedly differentiating and substituting, algebraically eliminating all components of the state until only a relation between y, u, and their derivatives remains, one can verify that (8) is equivalent to (3).

The solution of the top equation of (8) in the time domain is

$$\mathbf{x}(t) = e^{t\mathbf{A}}\mathbf{x}(0) + \int_0^t e^{(t-\tau)\mathbf{A}}\mathbf{b}u(\tau)\,d\tau \tag{9}$$

The matrix exponential is defined by the power series

$$e^{t\mathbf{A}} = \sum_{k=0}^{\infty} \frac{t^k}{k!}\mathbf{A}^k \tag{10}$$

The connection between state-space methods for time-invariant systems and the Laplace transform is made by noting the transform pair

$$\mathscr{L}\{e^{t\mathbf{A}}\} = (s\mathbf{I} - \mathbf{A})^{-1} \tag{11}$$

Substituting (9) into the second equation of (8) gives

$$y(t) = \mathbf{c}^T e^{t\mathbf{A}}\mathbf{x}(0) + \int_0^t \mathbf{c}^T e^{(t-\tau)\mathbf{A}}\mathbf{b}u(\tau)\,d\tau \tag{12}$$

Taking the Laplace transform of (12), using (11) and the rule for the transform of a convolution integral, gives

$$Y(s) = \mathbf{c}^T(s\mathbf{I} - \mathbf{A})^{-1}\mathbf{x}(0) + \mathbf{c}^T(s\mathbf{I} - \mathbf{A})^{-1}\mathbf{b}U(s) \tag{13}$$

The term $\mathbf{c}^T(s\mathbf{I} - \mathbf{A})^{-1}\mathbf{x}(0)$ represents the part of the output due to the initial state $\mathbf{x}(0)$, that is, due to the initial conditions. If these are zero, (13) reduces to (1), where we recognize

$$H(s) = \mathbf{c}^T(s\mathbf{I} - \mathbf{A})^{-1}\mathbf{b} \tag{14}$$

One advantage of state-space techniques is that once we have introduced the vector/matrix notation, we can extend it effortlessly to cover systems with multiple inputs and/or multiple outputs. Thus, if the input is a vector of dimension r and the output is a vector of dimension p, then (8) extends to

$$\begin{aligned} \dot{\mathbf{x}}(t) &= \mathbf{A}\mathbf{x}(t) + \mathbf{B}\mathbf{u}(t) \\ \mathbf{y}(t) &= \mathbf{C}\mathbf{x}(t) \end{aligned} \tag{15}$$

Here $\mathbf{B}$ is an $n \times r$ matrix and $\mathbf{C}$ is a $p \times n$ matrix. In the multiple input/multiple output case there is less advantage to the canonical form (4).

Time-Varying Systems

The reason for introducing state space in this chapter was to permit the discussion of time-varying systems. A standard model for a time-varying linear system is the extension of (15) to the case of time-varying parameter matrices:

$$\begin{aligned} \dot{\mathbf{x}}(t) &= \mathbf{A}(t)\mathbf{x}(t) + \mathbf{B}(t)\mathbf{u}(t) \\ \mathbf{y}(t) &= \mathbf{C}(t)\mathbf{x}(t) \end{aligned} \tag{16}$$

In this case, the starting time is no longer necessarily taken to be zero, but may be some arbitrary time t_0. When $\mathbf{A}(t)$ varies with time, the matrix exponential $e^{t\mathbf{A}}$ must be replaced by the transition matrix $\mathbf{\Phi}(t, t_0)$. In general, there is no analytical expression analogous to (10) for the transition matrix of a time-varying matrix.

The analog of equation (9), the input/state relation, for a multiple input time-varying system is

$$\mathbf{x}(t) = \mathbf{\Phi}(t, t_0)\mathbf{x}(t_0) + \int_{t_0}^{t} \mathbf{\Phi}(t, \tau)\mathbf{B}(\tau)\mathbf{u}(\tau)\, d\tau \tag{17}$$

In many applications of linear system theory to stochastic problems, one finds the model (16) being used with the deterministic input vector $\mathbf{u}(t)$ replaced by vector-valued white noise $\mathbf{V}(t)$. Vector-valued white noise is defined as a vector, each component of which is the white noise process discussed in connection with equations (45)–(47) of the previous chapter. Actually, this concept can be generalized in two ways, which is customarily done.

First, since we are now considering nonstationary processes, the white noise itself can be nonstationary. The way in which a sample of the noise at a given time correlates with a sample at another time is still represented by a δ-function as in equation (47) of Chapter 6, but the intensity of the noise process itself is allowed to vary with time, which means that the δ-function is multiplied by a time-varying amplitude (usually assumed to be a continuous function, or at worst, piecewise continuous).

Second, since we are talking of vector white noise, there may be correlation between one component of the vector and another. To represent this, the "time-varying amplitude" function just mentioned is allowed to be matrix-valued.

For full generality, the mean of the noise process may not be zero. Let us now put into equations what we just said in words. Here is a vector-valued white noise $\mathbf{V}(t)$. It is a Gaussian stochastic process having mean

$$E\{\mathbf{V}(t)\} = \mathbf{m}_V(t) \tag{18}$$

and autocovariance

$$\begin{aligned}\mathbf{C}_{VV}(t_1, t_2) &= E\{[\mathbf{V}(t_1) - \mathbf{m}_V(t_1)][\mathbf{V}(t_2) - \mathbf{m}_V(t_2)]^T\} \\ &= \mathbf{Q}(t_1)\delta(t_1 - t_2)\end{aligned} \tag{19}$$

If the dimension of $\mathbf{V}(t)$ is r, then $\mathbf{m}_V(t)$ is a vector of the same dimension, and $\mathbf{Q}(t)$ is a time-varying non-negative-definite $r \times r$ matrix.

When $\mathbf{V}(t)$ is used as input to the system (16), let $\mathbf{X}(t)$ denote the state process. Let the state at time t_0 be the Gaussian random vector $\mathbf{X}(t_0)$, having mean vector $\mathbf{m}_0$ and covariance matrix $\mathbf{R}_0$. Let us assume $\mathbf{X}(t_0)$ is independent of the process $\{\mathbf{V}(t)\}$. The state–output relation (17) in this case becomes

$$\mathbf{X}(t) = \mathbf{\Phi}(t, t_0)\mathbf{X}(t_0) + \int_{t_0}^{t} \mathbf{\Phi}(t, \tau)\mathbf{B}(\tau)\mathbf{V}(\tau)\, d\tau \tag{20}$$

Taking the expected value of (20) yields

$$E\{\mathbf{X}(t)\} = \mathbf{m}_X(t) = \mathbf{\Phi}(t, t_0)\mathbf{m}_0 + \int_{t_0}^{t} \mathbf{\Phi}(t, \tau)\mathbf{B}(\tau)\mathbf{m}_V(\tau)\, d\tau \tag{21}$$

We calculate the autocovariance matrix of the state process $\{\mathbf{X}(t)\}$, recalling the agreements made above:

$$
\begin{aligned}
&\mathbf{C}_{XX}(t_1, t_2) \\
&\quad = E\{[\mathbf{X}(t_1) - \mathbf{m}_X(t_1)][\mathbf{X}(t_2) - \mathbf{m}_X(t_2)]^T\} \\
&\quad = E\Bigg\{\bigg[\mathbf{\Phi}(t_1, t_0)(\mathbf{X}(t_0) - \mathbf{m}_0) + \int_{t_0}^{t_1}\mathbf{\Phi}(t_1, \tau)\mathbf{B}(\tau)(\mathbf{V}(\tau) - \mathbf{m}_V(\tau))\, d\tau\bigg] \\
&\qquad \times\bigg[\mathbf{\Phi}(t_2, t_0)(\mathbf{X}(t_0) - \mathbf{m}_0) + \int_{t_0}^{t_2}\mathbf{\Phi}(t_2, \sigma)\mathbf{B}(\sigma)(\mathbf{V}(\sigma) - \mathbf{m}_V(\sigma))\, d\sigma\bigg]^T\Bigg\} \\
&\quad = \mathbf{\Phi}(t_1, t_0)\mathbf{E}\left\{(\mathbf{X}(t_0) - \mathbf{m}_0)(\mathbf{X}(t_0) - \mathbf{m}_0)^T\right\}\mathbf{\Phi}^T(t_2, t_0) \\
&\qquad + \int_{t_0}^{t_1}\int_{t_0}^{t_2}\mathbf{\Phi}(t_1, \tau)\mathbf{B}(\tau)E\left\{(\mathbf{V}(\tau) - \mathbf{m}_V(\tau))(\mathbf{V}(\sigma) - \mathbf{m}_V(\sigma))^T\right\} \\
&\qquad\qquad \times \mathbf{B}^T(\sigma)\mathbf{\Phi}^T(t_2, \sigma)\, d\tau\, d\sigma \\
&\quad = \mathbf{\Phi}(t_1, t_0)\mathbf{R}_0\mathbf{\Phi}^T(t_2, t_0) \\
&\qquad + \int_{t_0}^{t_1}\int_{t_0}^{t_2}\mathbf{\Phi}(t_1, \tau)\mathbf{B}(\tau)\mathbf{Q}(\tau)\delta(\tau - \sigma)\mathbf{B}^T(\sigma)\mathbf{\Phi}^T(t_2, \sigma)\, d\tau\, d\sigma
\end{aligned}
\tag{22}
$$

The reader should study the string of equations (22). He or she should be able to explain what each line says and to justify how it was obtained.

It still remains to evaluate the integral term in the last line. To do this, assume first that $t_1 > t_2$, $t_0 \le \tau \le t_1$, and $t_0 \le \sigma \le t_2$, and do the integration first with respect to τ, then with respect to σ. In that case, whatever the value of σ, τ will sweep past it and "fire" the δ-function. Therefore, for $t_1 > t_2$ we have

$$
\begin{aligned}
&\int_{t_0}^{t_1}\int_{t_0}^{t_2}\mathbf{\Phi}(t_1, \tau)\mathbf{B}(\tau)\mathbf{Q}(\tau)\delta(\tau - \sigma)\mathbf{B}^T(\sigma)\mathbf{\Phi}^T(t_2, \sigma)\, d\tau\, d\sigma \\
&\qquad = \int_{t_0}^{t_2}\mathbf{\Phi}(t_1, \sigma)\mathbf{B}(\sigma)\mathbf{Q}(\sigma)\mathbf{B}^T(\sigma)\mathbf{\Phi}^T(t_2, \sigma)\, d\sigma
\end{aligned}
\tag{23}
$$

The transition matrix necessarily possesses the transition property, which we require in two situations:

$$
\begin{aligned}
\mathbf{\Phi}(t_1, t_0) &= \mathbf{\Phi}(t_1, t_2)\mathbf{\Phi}(t_2, t_0) \\
\mathbf{\Phi}(t_1, \sigma) &= \mathbf{\Phi}(t_1, t_2)\mathbf{\Phi}(t_2, \sigma)
\end{aligned}
\tag{24}
$$

Using (23) and (24), (22) may be written

$$\mathbf{C}_{XX}(t_1, t_2) = \mathbf{\Phi}(t_1, t_2)\mathbf{C}_{XX}(t_2, t_2) \quad \text{for } t_1 > t_2 \tag{25}$$

where

$$\mathbf{C}_{XX}(t_2, t_2) = \mathbf{\Phi}(t_2, t_0)\mathbf{R}_0\mathbf{\Phi}^T(t_2, t_0) + \int_{t_0}^{t_2} \mathbf{\Phi}(t_2, \sigma)\mathbf{B}(\sigma)\mathbf{Q}(\sigma)\mathbf{B}^T(\sigma)\mathbf{\Phi}^T(t_2, \sigma)\, d\sigma \tag{26}$$

For the case $t_1 < t_2$, repeating the entire procedure of evaluating the integral with appropriate modifications leads to

$$\mathbf{C}_{XX}(t_1, t_2) = \mathbf{C}_{XX}(t_1, t_1)\mathbf{\Phi}^T(t_2, t_1) \quad \text{for } t_1 < t_2 \tag{27}$$

where

$$\mathbf{C}_{XX}(t_1, t_1) = \mathbf{\Phi}(t_1, t_0)\mathbf{R}_0\mathbf{\Phi}^T(t_1, t_0) + \int_{t_0}^{t_1} \mathbf{\Phi}(t_1, \tau)\mathbf{B}(\tau)\mathbf{Q}(\tau)\mathbf{B}^T(\tau)\mathbf{\Phi}^T(t_1, \tau)\, d\tau \tag{28}$$

Equations (25)–(28) are the main results for the autocovariance matrix $\mathbf{C}_{XX}(t_1, t_2)$ of the $\{\mathbf{X}(t)\}$ process generated by (20). In many applications, however, it is not the $\{\mathbf{X}(t)\}$ process that is being studied, but rather the system output as given by the second of equations (16). Since $\mathbf{X}(t)$ is now random, we need to denote this output by a capital letter also, thus:

$$\mathbf{Y}(t) = \mathbf{C}(t)\mathbf{X}(t) \tag{29}$$

Here $\mathbf{C}(t)$ is a given, deterministic, time-varying matrix.

From (29) it follows at once that

$$\mathbf{m}_Y(t) = \mathbf{C}(t)\mathbf{m}_X(t) \tag{30}$$

$$\mathbf{C}_{YY}(t_1, t_2) = \mathbf{C}(t_1)\mathbf{C}_{XX}(t_1, t_2)\mathbf{C}^T(t_2) \tag{31}$$

We have discussed the consequences of stochasticizing the model (16) by using white noise $\{\mathbf{V}(t)\}$ as the input. By proper choice of the matrices $\mathbf{A}$, $\mathbf{B}$, and $\mathbf{C}$, a wide variety of mean and autocovariance functions for the output process $\{\mathbf{Y}(t)\}$ can be obtained, as given by (30) and (31). Some of the further possibilities are illustrated by the problems at the end of this chapter.

In a number of applications, the process $\{\mathbf{Y}(t)\}$ is used to represent a signal that is transmitted over some communication channel, where it suffers the addition of some interference. The interference is often modeled as white noise, usually assumed to be independent of the $\{\mathbf{Y}(t)\}$ process. Thus, the output of the channel would be $\mathbf{Z}(t)$, where

$$\mathbf{Z}(t) = \mathbf{Y}(t) + \mathbf{N}(t) \tag{32}$$

and $\{\mathbf{N}(t)\}$ is continuous-time vector-valued white noise, with mean $\mathbf{m}_N(t)$ and covariance $\mathbf{R}(t_1)\delta(t_1 - t_2)$.

Since we assume $\{\mathbf{Y}(t)\}$ and $\{\mathbf{N}(t)\}$ to be independent, it follows at once that

$$E\{\mathbf{Z}(t)\} = \mathbf{m}_Z(t) = \mathbf{m}_Y(t) + \mathbf{m}_N(t) \tag{33}$$

$$\begin{aligned} &E\{[\mathbf{Z}(t_1) - \mathbf{m}_Z(t_1)][\mathbf{Z}(t_2) - \mathbf{m}_Z(t_2)]^T\} \\ &\quad = \mathbf{C}_{ZZ}(t_1, t_2) = \mathbf{C}_{YY}(t_1, t_2) + \mathbf{R}(t_1)\delta(t_1 - t_2) \end{aligned} \tag{34}$$

Problems

1. For $\mathbf{m}_X(t)$ as defined in (21), find a deterministic differential equation with appropriate initial condition, of which $\mathbf{m}_X(t)$ is the solution.

2. For $\mathbf{C}_{XX}(t_1, t_2)$ as defined in (22):

a. Show that for $t_1 > t_2$ the following differential equation holds:

$$\frac{\partial}{\partial t_1}\mathbf{C}_{XX}(t_1, t_2) = \mathbf{A}(t_1)\mathbf{C}_{XX}(t_1, t_2)$$

b. Show that for $t_1 < t_2$ the following differential equation holds:

$$\frac{\partial}{\partial t_2}\mathbf{C}_{XX}(t_1, t_2) = \mathbf{C}_{XX}(t_1, t_2)\mathbf{A}^T(t_2)$$

c. Find the differential equation obeyed by $\mathbf{C}_{XX}(t_1, t_1)$, the solution of which is given by (28).

In parts a and b, $\mathbf{A}$ is the matrix from (16).

3. This problem explores some of what happens when a stochastic process other than white noise is used as input to (16). Let $\{\mathbf{W}(t)\}$ be a Gaussian process with mean $\mathbf{m}_W(t)$ and autocovariance $\mathbf{C}_{WW}(t_1, t_2)$.

Consider the state–input equation

$$\dot{\mathbf{X}}(t) = \mathbf{A}(t)\mathbf{X}(t) + \mathbf{B}(t)\mathbf{W}(t)$$

a. Find an explicit expression for the mean $\mathbf{m}_X(t)$.

b. Find the cross-covariance

$$\mathbf{C}_{XW}(t_1, t_2) = E\{[\mathbf{X}(t_1) - \mathbf{m}_X(t_1)][\mathbf{W}(t_2) - \mathbf{m}_W(t_2)]^T\}$$

4. Suppose the same white noise process $\{\mathbf{V}(t)\}$, with mean $\mathbf{m}_V(t)$ and autocovariance $\mathbf{Q}(t_1)\delta(t_1 - t_2)$, is used as the input to two different systems,

$$\dot{\mathbf{X}}_1(t) = \mathbf{A}_1(t)\mathbf{X}_1(t) + \mathbf{B}_1(t)\mathbf{V}(t)$$

$$\dot{\mathbf{X}}_2(t) = \mathbf{A}_2(t)\mathbf{X}_2(t) + \mathbf{B}_2(t)\mathbf{V}(t)$$

with associated outputs

$$\mathbf{Y}_1(t) = \mathbf{C}_1(t)\mathbf{X}_1(t)$$

$$\mathbf{Y}_2(t) = \mathbf{C}_2(t)\mathbf{X}_2(t)$$

Find the cross-covariance $\mathbf{C}_{Y_1Y_2}(t_1, t_2)$.

5. Suppose a time-varying linear system has a scalar input $V(t)$ and scalar output $Y(t)$, with input–output relation

$$Y(t) = \int_{t_0}^{t} W(t, \tau)V(\tau)\, d\tau$$

where $W(t, \tau)$ is a given deterministic function of two arguments, called the system weighting pattern.

a. If $V(t)$ is unit WGN with mean zero, and if the system is assumed to be internally at rest at the starting time t_0, show that the resulting autocovariance function for the $\{Y(t)\}$ process is

$$c_{YY}(t_1, t_2) = \int_{t_0}^{\min\{t_1, t_2\}} W(t_1, \tau)W(t_2, \tau)\, d\tau$$

Note: $\min\{t_1, t_2\}$ means the smaller of the two numbers t_1, t_2.

b. Suppose the scalar function $W(t, \tau)$ can be factored as the scalar product of two n-vectors $\mathbf{c}(t)$ and $\mathbf{b}(\tau)$, where $\mathbf{c}(t)$ is a function only of t and $\mathbf{b}(\tau)$ is a function only of τ, thus:

$$W(t, \tau) = \mathbf{c}^T(t)\mathbf{b}(\tau)$$

In that case, show that the system can be realized as a set of n time-varying gains all having the same input $V(t)$, followed by n integrators, one in each channel, followed by n more time-varying gains all having the same output $Y(t)$. In equation form, we have

$$\dot{\mathbf{X}}(t) = \mathbf{b}(t)V(t)$$

$$Y(t) = \mathbf{c}^T(t)\mathbf{X}(t)$$

Chapter 8

Additional Topics in the Study of Continuous-Time Processes

Introduction

In the previous chapters we have discussed at some length the techniques of modeling a stochastic process as the output of a linear system excited by white noise. Such a model often expedites analyses or calculations concerning the stochastic process, because the probabilistic character of the problem is transferred to the white noise. Calculations of correlations and similar quantities are facilitated by this maneuver because of the independence properties of white noise.

In this chapter we examine another representation of a stochastic process, the advantage of which basically rests on the same principle. The idea is to represent each sample function of the process by a series expansion in known, deterministic functions with random coefficients. We will be totally concerned with second-order features of the process here, so the property of independence will be replaced by that of *orthogonality*. In Chapter 1 we discussed briefly the Hilbert space of second-order random variables and introduced the interpretation of the covariance of two zero-mean random variables as an inner product in this Hilbert space.

The principle of orthogonality will also play a fundamental role in regard to the deterministic time functions used as the basis of the expansion. In order to set the stage for that aspect of our presentation, it is necessary now to discuss briefly the properties of a Hilbert space whose elements are functions of time. We will then take up the subject of integral equations and show how such equations can be regarded in terms of operators and vectors in Hilbert space. That will then put us in a position to obtain the desired orthogonal expansion of a second-order stochastic process.

Hilbert Space of Time Functions on an Interval

The material we are about to present has its rigorous foundations in the mathematical discipline called real analysis, which is based on measure theory. Since that background is not required of the reader of this book, it is not possible to be totally precise without creating obstacles to the smooth flow of those ideas that we *do* want to get across. By way of apology to the initiated, we offer a brief disclaimer here, to the effect that we are not aiming at strict semantic precision, but rather, at developing a facility on the part of the reader for dealing with these various mathematical objects in practical applications.

Let us now proceed. Hilbert space is a vector space having an infinite number of dimensions, and is further endowed with a set of properties that make it the most natural generalization of familiar finite-dimensional Euclidean space. Although it can be defined and studied in a totally abstract way, without ever even saying that one concrete example is a set of real-valued functions, it is adequate here if the reader thinks the set of functions is the same thing as the space. Therefore, consider the set of all real-valued functions of the real variable t, $0 \leqslant t \leqslant T$. The Hilbert space in which we are interested is the subset of those functions that also meet the condition

$$\int_0^T f^2(t)\,dt < \infty \tag{1}$$

It is customary to denote this particular Hilbert space by $L_2[0, T]$.

The mathematical properties which make Hilbert space distinct from other infinite-dimensional vector spaces are its norm and its inner product. If f is a function in $L_2[0, T]$, its *norm*, denoted $\|f\|$, is defined by

$$\|f\| = \left[\int_0^T f^2(t)\,dt\right]^{1/2} \tag{2}$$

If f and g are two functions in $L_2[0, T]$, their inner product, $\langle f, g\rangle$, is defined as

$$\langle f, g\rangle = \int_0^T f(t)g(t)\,dt \tag{3}$$

It is essential, in order to be a Hilbert space, that the inner product and the norm are related as indicated in (2) and (3). Without using the integrals, we may write directly

$$\|f\| = \langle f, f \rangle^{1/2} \tag{4}$$

$$\langle f, g \rangle = \tfrac{1}{4}[\|f + g\|^2 - \|f - g\|^2] \tag{5}$$

Probably the reader became familiar with the properties of Euclidean vector space from his studies of linear algebra, where a set of n simultaneous equations in n unknowns is represented by

$$\mathbf{Ax} = \mathbf{y} \tag{6}$$

where $\mathbf{A}$ is an $n \times n$ matrix of coefficients, $\mathbf{x}$ is the n-dimensional vector of unknowns, and $\mathbf{y}$ is a given n-dimensional vector. The Hilbert space generalization of (6) is the operator equation

$$Lf = g \tag{7}$$

The only kind of operator we will consider here is the *integral operator*, which is characterized by a kernel function $K(s, t)$. Here, K is a real-valued function defined on the square $0 \leqslant s \leqslant T$, $0 \leqslant t \leqslant T$. The abstract expression (7) represents the explicit integral equation

$$\int_0^T K(s, t) f(t)\, dt = g(s) \tag{8}$$

In analogy with matrices, a very valuable avenue for the study of the structure of operators in Hilbert space is spectral theory. For matrices, this begins with consideration of the eigenvalue/eigenvector problem

$$\mathbf{Ax} = \lambda \mathbf{x} \tag{9}$$

which we assume is quite familiar to the reader. The general spectral theory for linear operators in Hilbert space is much more complicated than in the finite-dimensional case, but there is one class of operators for which the spectral theory most resembles the finite-dimensional results. This class is known as the class of compact, self-adjoint operators. Fortunately for workers in stochastic process theory, the autocovariance function $c(t, s)$ for a continuous-time stochastic process as defined in equation (2) of Chapter 6, when used as the kernel function $K(t, s)$ of an integral operator such as (8) above (on a finite interval $0 \leqslant t \leqslant T$), turns out always to yield a compact, self-adjoint operator. We can therefore confine our discussion to this class of operators. After a discussion of the properties of compact, self-adjoint operators, we will proceed directly to show the application of this theory to stochastic processes.

DEFINITION. A linear operator L which maps a Hilbert space $\mathscr{H}$ into itself is called *self-adjoint* if and only if, for all pairs of elements f, g in $\mathscr{H}$ it holds that

$$\langle f, Lg \rangle = \langle Lf, g \rangle \tag{10}$$

Here the definition is given abstractly in terms of the inner product. For integral operators characterized by (8) in the Hilbert space $L_2[0, T]$, the corresponding condition on the kernel function $K(s, t)$ is simply

$$K(s, t) = K(t, s) \tag{11}$$

The student is familiar with the concept of a *basis* for a vector space from his or her studies in linear algebra. For a finite-dimensional space, the number of basis elements equals the dimension of the space. For an infinite-dimensional space such as $L_2[0, T]$, an infinite number of basis elements are required. There is a complication here, in that the issue of "orders of infinity" must be considered. An infinite set of objects is called "countably infinite" if and only if its members can be put in one-to-one correspondence with the positive integers. If this cannot be done, then the set is of a higher order of infinity than "countable."

It turns out that the number of basis elements required for $L_2[0, T]$ is countable. In fact, one example of a basis is the familiar Fourier sine series set of elements:

$$\sin\left(\frac{n\pi t}{T}\right), \qquad n = 1, 2, \ldots$$

Let $\{e_n\}$ be a basis for a Hilbert space H. Just as in the finite-dimensional case, we say the basis is *orthonormal* (O.N.) if its satisfies

$$\langle e_n, e_m \rangle = \begin{cases} 0, & n \neq m \\ 1, & n = m \end{cases} \tag{12}$$

The precise definition of a *compact* operator requires more discussion of modes of convergence in Hilbert space than we wish to provide in this book. However, a useful pragmatic definition that is sufficient for compactness is: L is compact if

$$\| Le_n \| \to 0 \quad \text{as } n \to \infty \tag{13}$$

for every orthonormal basis $\{e_n\}$.

The set of eigenvalues of a finite-dimensional operator is called its *spectrum*. For operators on an infinite-dimensional space, the situation is more complicated. For example, it is possible for an operator to have a continuous spectrum, the individual points of which are not considered eigenvalues. Even with a discrete spectrum, however, there is a complication that we illustrate with this example. An operator might have the set of eigenvalues

$$\lambda_n = \frac{1}{n}, \qquad n = 1, 2, \ldots \tag{14}$$

The spectrum is defined to consist of these distinct points and the point zero, because zero is the limit of the sequence. However, zero itself might not be an eigenvalue. Think of an infinite diagonal matrix with the elements (14) on the diagonal. The operator represented by this matrix is nonsingular, because no nonzero element in the space is mapped into the origin, that is, the null space of the operator contains only the origin. Therefore zero itself is not an eigenvalue. However, by convention it *is* considered to belong to the spectrum of the operator.

A compact operator that is not self-adjoint could have a spectrum that consists only of the single point zero, but it is possible for that point not to be an eigenvalue. We could illustrate this again with an infinite matrix. Use the elements (14) again, but put them on the diagonal just below the main diagonal. For the pertinent orthonormal basis, this operator is characterized by

$$Le_n = \frac{1}{n} e_{n+1}, \qquad n = 1, 2, \ldots \tag{15}$$

This operator has the property just described.

We hope that these examples will make the following theorems intelligible.

Theorem. The spectrum of a compact operator is a countable set, the only possible limit point of which is zero. Every nonzero point of the spectrum is an eigenvalue.

As the example shows, it might be that a compact operator has no eigenvalues at all. When the self-adjointness property is included, then this peculiarity cannot occur.

Theorem. Every compact self-adjoint operator mapping a Hilbert space $\mathscr{H}$ into itself has at least one eigenvalue.

Theorem. The eigenvectors of a compact, self-adjoint operator are mutually orthogonal.

For proofs of all of these theorems, see Balakrishnan (1981).

The Karhunen–Loeve Expansion

Now consider an integral operator having a covariance kernel $c(t, s)$ as defined by equation (2) of Chapter 6, associated with a second-order process $\{X(\cdot)\}$. This operator is positive definite in the sense of the discussion following equation (4) of that chapter. It turns out (Bochner's theorem again) that it is also positive definite in the sense that all of its eigenvalues are positive.

The objects that correspond to eigenvectors of a matrix are called, in the case of an integral operator, eigenfunctions. Let $\{\lambda_n\}$ be the set of nonzero eigenvalues and $\{\phi_n\}$ the corresponding set of normalized eigenfunctions. Then we have $\lambda_n > 0$ for every n, and we will assume that all eigenvalues are simple, that is, of multiplicity one. Then to each λ_n there corresponds a function $\phi_n(t)$, $0 \leqslant t \leqslant T$, such that

$$\int_0^T c(s, t)\phi_n(t)\, dt = \lambda_n \phi_n(s) \tag{16}$$

These functions are an orthonormal set:

$$\int_0^T \phi_n(t)\phi_m(t)\, dt = \begin{cases} 1, & n = m \\ 0, & n \neq m \end{cases} \tag{17}$$

Let us assume further that $c(t, s)$ is jointly continuous in both variables. This puts us in a position to make use of a theorem called *Mercer's theorem*, which states that $c(t, s)$ has the representation

$$c(t, s) = \sum_{n=1}^{\infty} \lambda_n \phi_n(t)\phi_n(s) \tag{18}$$

(see Balakrishnan, 1981, pp. 125–130; Courant-Hilbert, 1953, pp. 138–140). The eigenfunctions ϕ_n are continuous on $[0, T]$ and the expansion (18) converges uniformly on $[0, T] \times [0, T]$.

The idea we now want to introduce is that of using the set of eigenfunctions $\{\phi_n\}$ as a basis for expanding the sample functions $X(\cdot)$ of the stochastic process $\{X(\cdot)\}$. This is the concept toward which all of the preceding discussion has been directed.

The property that makes the expansion of a stochastic process more sophisticated than the Fourier series or other orthonormal expansion of a deterministic function is that the coefficients are random variables. Specifically, define

$$\alpha_n = \int_0^T \phi_n(t) X(t)\, dt, \qquad n = 1, 2, \ldots \tag{19}$$

Then, since $X(\cdot)$ is random, so are all the α_n.

Equation (19) now implies the expansion

$$X(t) = \sum_{n=1}^{\infty} \alpha_n \phi_n(t) \tag{20}$$

So far, the form of the expansion (20) with random coefficients is what we would obtain if we used any orthonormal basis. The idea of representing a stochastic process by expanding it with respect to an orthonormal set of functions using random coefficients is a major idea, and much of the rest of this chapter will be devoted to it. However, there is a second level of sophistication here. Let us investigate the consequences of our particular choice of O.N. basis as the eigenfunctions of the covariance operator. From (19), we have

$$\begin{aligned} E\{\alpha_n \alpha_m\} &= E\left\{\left[\int_0^T \phi_n(t) X(t)\, dt\right]\left[\int_0^T \phi_m(s) X(s)\, ds\right]\right\} \\ &= E\left\{\int_0^T \int_0^T \phi_n(t) \phi_m(s) X(t) X(s)\, dt\, ds\right\} \\ &= \int_0^T \int_0^T \phi_n(t) \phi_m(s) E\{X(t) X(s)\}\, dt\, ds \end{aligned} \tag{21}$$

Let us assume henceforth that the mean $\mu(t)$ of our process is zero. If it is not, then (20) would have to be replaced by

$$X(t) = \mu(t) + \sum_{n=1}^{\infty} \alpha_n \phi_n(t) \tag{22}$$

In (21), it is Fubini's theorem of real analysis that justifies the interchange of expectation and time integration. With our assumption of zero mean,

$$E\{X(t) X(s)\} = c(t, s) \tag{23}$$

Therefore (21) becomes

$$E\{\alpha_n \alpha_m\} = \int_0^T \int_0^T \phi_n(t)\phi_m(s)c(t, s)\, dt\, ds \tag{24}$$

Using (16) and (17) in (24) yields

$$E\{\alpha_n \alpha_m\} = \begin{cases} \lambda_n, & n = m \\ 0, & n \neq m \end{cases} \tag{25}$$

Therefore, the consequence of our particular choice of orthonormal basis is that the expansion coefficients are themselves statistically orthogonal. That is, the coefficients are orthogonal vectors in the Hilbert space of second-order random variables discussed in Chapter 1. The purpose of this expansion (20) is to view a second-order stochastic process in the context of two Hilbert spaces: a Hilbert space of second-order random variables for the coefficients and a Hilbert space of time functions for the basis functions. Such an expansion is called a *Karhunen–Loeve expansion*.

Since the mean of the $\{X(\cdot)\}$ process is zero, the mean of all the coefficients α_n is also zero. From (25), we see that, in particular,

$$E\{\alpha_n^2\} = \lambda_n \tag{26}$$

so that the variance of the coefficient α_n is the corresponding eigenvalue.

The need for Mercer's theorem and the expansion (18) is to be able to prove that the expansion (20) does indeed converge statistically in the mean-square sense, pointwise in t. Let us do this now. For each value of t we have

$$\begin{aligned} &E\left\{\left[X(t) - \sum_{n=1}^{N} \alpha_n \phi_n(t)\right]^2\right\} \\ &\quad = E\left\{X^2(t) - 2\sum_{n=1}^{N} \alpha_n X(t)\phi_n(t) + \left[\sum_{n=1}^{N} \alpha_n \phi_n(t)\right]^2\right\} \\ &\quad = c(t, t) - 2\sum_{n=1}^{N} E\{\alpha_n X(t)\}\phi_n(t) \\ &\qquad + E\left\{\sum_{n=1}^{N}\sum_{m=1}^{N} \alpha_n \alpha_m \phi_n(t)\phi_m(t)\right\} \end{aligned} \tag{27}$$

Multiply both sides of (19) by $X(s)$ and take the expected value.

$$E\{\alpha_n X(s)\} = \int_0^T \phi_n(t) E\{X(t)X(s)\}\,dt$$

$$= \int_0^T \phi_n(t) c(t, s)\,dt = \lambda_n \phi_n(s) \tag{28}$$

where we have used (16).

Interchange the order of expectation and summation in the last term in (27), then use (25). With these changes, (27) becomes

$$c(t, t) - 2\sum_{n=1}^{N} \lambda_n \phi_n^2(t) + \sum_{n=1}^{N} \lambda_n \phi_n^2(t) = c(t, t) - \sum_{n=1}^{N} \lambda_n \phi_n^2(t) \tag{29}$$

But as $N \to \infty$, the summation in (29) converges to $c(t, t)$; this is what (18) says when $s = t$.

Therefore, for a stochastic process $\{X(\cdot)\}$ having a covariance $c(t, s)$ which is jointly continuous in both variables, the Karhunen–Loeve expansion converges in the mean square sense statistically, pointwise in t. This requirement on $c(t, s)$ is equivalent to mean-squared continuity of the process $\{X(\cdot)\}$ at *every* point t_0, $0 \leqslant t_0 \leqslant T$, as indicated by equation (6) of Chapter 6 (see Wong and Hajek, 1985, pp. 77–78).

If $c(t, s)$ is not jointly continuous, so that Mercer's theorem is not available, then instead of convergence pointwise in t we get convergence in the sense of $L_2[0, T]$; that is, the expansion (20) holds in the sense that

$$\lim_{N \to \infty} E\left\{\int_0^T \left[X(t) - \sum_{n=1}^{N} \alpha_n \phi_n(t)\right]^2 dt\right\} = 0 \tag{30}$$

The only condition required to obtain (30) is

$$E\left\{\int_0^T X^2(t)\,dt\right\} < \infty \tag{31}$$

which, by interchanging the order of integration and expectation, is equivalent to

$$\int_0^T c(t, t)\,dt < \infty \tag{32}$$

In terms of the eigenvalues of the covariance operator, (32) in turn is equivalent to

$$\sum_{n=1}^{\infty} \lambda_n < \infty \tag{33}$$

(see Balakrishnan, 1981, Theorem 3.4.4).

Example: Karhunen–Loeve Expansion of Brownian Motion

Although to be completely rigorous one should define the Brownian motion process, also called the Wiener process, directly from basic principles, we will introduce it here in terms of the white noise process $\{V(\cdot)\}$ defined in connection with (45)–(47) of Chapter 6. Brownian motion $\{B(\cdot)\}$ is simply the time integral of white noise:

$$B(t) = \int_0^t V(\tau)\, d\tau \tag{34}$$

It is customary to take the white noise with zero mean, so that Brownian motion has the same property.

Let us find the autocovariance $c_{BB}(t, s)$ of Brownian motion:

$$\begin{aligned} c_{BB}(t, s) &= E\{B(t)B(s)\} \\ &= E\left\{\left[\int_0^t V(\tau)\, d\tau\right]\left[\int_0^s V(\sigma)\, d\sigma\right]\right\} \\ &= E\left\{\int_0^t \int_0^s V(\tau)V(\sigma)\, d\tau\, d\sigma\right\} \\ &= \int_0^t \int_0^s E\{V(\tau)V(\sigma)\}\, d\tau\, d\sigma \\ &= \int_0^t \int_0^s \delta(\tau - \sigma)\, d\tau\, d\sigma \\ &= \min(t, s) = \begin{cases} s, & 0 \leqslant s \leqslant t \leqslant 1 \\ t, & 0 \leqslant t \leqslant s \leqslant 1 \end{cases} \end{aligned} \tag{35}$$

The alert reader will recognize that this calculation resembles Equation (22) of Chapter 7.

The Brownian motion process has so many curious properties that it is almost a whole field of study for mathematicians all by itself. A description of these would take us too far afield. See Wong and Hajek (1985), where further references can be found. Our focus right now is to find the Karhunen–Loeve expansion for $\{B(\cdot)\}$.

We begin with (16), by finding the eigenvalues and eigenfunctions. For the covariance (35), (16) becomes

$$\int_0^s t\phi_n(t)\,dt + s\int_s^1 \phi_n(t)\,dt = \lambda_n\phi_n(s) \tag{36}$$

Simply differentiate both sides of (36) with respect to s once to obtain

$$\lambda_n \frac{d\phi_n(s)}{ds} = \int_s^1 \phi_n(t)\,dt \tag{37}$$

Differentiating a second time yields

$$\lambda_n \frac{d^2\phi_n(s)}{ds^2} = -\phi_n(s) \tag{38}$$

Setting $s = 0$ in (36) yields $\phi_n(0) = 0$, while setting $s = 1$ in (37) yields $\phi_n'(1) = 0$. Therefore, our eigenvalues and eigenfunctions are solutions to the two-point boundary value problem

$$\phi_n''(s) + \frac{1}{\lambda_n}\phi_n(s) = 0$$

$$\phi_n(0) = \phi_n'(1) = 0 \tag{39}$$

The orthonormalized eigenfunctions are

$$\phi_n(s) = \sqrt{2}\sin\frac{(2n-1)\pi s}{2} \qquad n = 1, 2, \ldots \tag{40}$$

The corresponding eigenvalues are

$$\lambda_n = \sqrt{\frac{2}{(2n-1)\pi}} \tag{41}$$

The Mercer's theorem representation of $c_{BB}(t, s)$ is, therefore, by (18),

$$c_{BB}(t,s) = 2\sqrt{\frac{2}{\pi}}\sum_{n=1}^{\infty}\frac{\sin\frac{1}{2}(2n-1)\pi t\,\sin\frac{1}{2}(2n-1)\pi s}{\sqrt{2n-1}} \tag{42}$$

From (20), Brownian motion has the Karhunen–Loeve representation

$$B(t) = \sum_{N=1}^{\infty} \alpha_n \sqrt{2} \sin \frac{(2n-1)\pi t}{2} \tag{43}$$

where the $\{\alpha_n\}$ are a set of mutually orthogonal, zero mean random variables. It follows from (26) and (41) that their variance is

$$E\{\alpha_n^2\} = \lambda_n = \sqrt{\frac{2}{(2n-1)\pi}} \tag{44}$$

Actually, to be precise, we should specify the probability distribution of the random variables $\{\alpha_n\}$. The Brownian motion process is strictly defined to be a Gaussian process. Up to this point we have used only second-order properties. However, to obtain true Brownian motion, we need to specify that the $\{\alpha_n\}$ all have Gaussian distributions with zero mean and variances given by (44).

Now, if r.v.'s are both zero mean and orthogonal, then they are uncorrelated, If Gaussian r.v.'s are uncorrelated, then they are *independent*. Thus, the $\{\alpha_n\}$ in this case are mutually independent.

Other Orthogonal Expansions

We may uniquely characterize the Karhunen–Loeve expansion as the expansion of the form (20) in which *both*

1. the $\{\phi_n\}$ are O.N. in $L_2[0, T$,
2. the $\{\alpha_n\}$ are statistically orthogonal.

There is an infinite variety of possible expansions of the form (20) if we give up one or the other of these requirements. In practice, sometimes it is worthwhile to give up property 2 in order to be able to use a standard set of basis functions, rather than having to find the eigenfunctions of the covariance operator. One standard set is the trigonometric functions, that is, simply use a Fourier series expansion. Although Fourier series is introduced in the context of periodic functions on an infinite time interval, of course if we confine our attention to a single finite interval $[0, T]$, then periodicity is no longer an issue.

In what follows it will be a great convenience to allow the stochastic processes, the expansion coefficients, and the basis functions all to be complex-valued. Since these are all complex-*valued* functions of a *real argument* (either time or index of summation), this does not open a new chapter of mathematics, but merely makes a notational innovation for convenience. The complex conjugate of any quantity will be denoted by a superscript asterisk. We will indicate any necessary modifications in our definitions of such things as auto- and cross-covariance functions, inner products, and so on, as we proceed.

We consider now an arbitrary complex-valued second-order process $\{X(\cdot)\}$ on the interval $[0, T]$. Its mean function $\mu_X(t)$, is therefore, complex-valued:

$$\mu_X(t) = E\{X(t)\} \tag{45}$$

Its autocovariance $c_{XX}(t, s)$ is also complex-valued, and here is where one of our former definitions gets generalized:

$$c_{XX}(t, s) = E\{[X(t) - \mu_X(t)][X(s) - \mu_X(s)]^*\} \tag{46}$$

Now, the symmetry property becomes

$$c_{XX}(t, s) = c_{XX}^*(s, t) \tag{47}$$

The Fourier series representation for $\{X(\cdot)\}$ is the same as for deterministic functions, except that the expansion coefficients are random variables:

$$X(t) = \sum_{k=-\infty}^{\infty} \zeta_k e^{ik\omega_0 t} \tag{48}$$

where $\omega_0 = 2\pi/T$, and the ζ_k are given by

$$\zeta_k = \frac{1}{T}\int_0^T X(t)e^{-ik\omega_0 t}\,dt \tag{49}$$

Since these basis functions were not chosen by the Karhunen–Loeve procedure, the coefficients are correlated. Taking expected values of both sides of (49),

$$\bar{\zeta}_k = E\{\zeta_k\} = \frac{1}{T}\int_0^T \mu_X(t)e^{-ik\omega_0 t}\,dt \tag{50}$$

The cross-covariance between ζ_k and ζ_l is

$$E\{[\zeta_k - \bar{\zeta}_k][\zeta_l - \bar{\zeta}_l]^*\} = \gamma_{kl}$$

$$= \frac{1}{T^2}\int_0^T\int_0^T c_{XX}(t, s)e^{-ik\omega_0 t}e^{il\omega_0 s}\,dt\,ds \tag{51}$$

The significance of the coefficients γ_{kl} is that they are the coefficients in the two-dimensional Fourier series representation of $c_{XX}(t, s)$ on the square $[0, T] \times [0, T]$.

$$c_{XX}(t, s) = \sum_{k=-\infty}^{\infty} \sum_{l=-\infty}^{\infty} \gamma_{kl} e^{ik\omega_0 t} e^{-il\omega_0 s} \tag{52}$$

Equation (52) should be compared to (18), which is the corresponding result for the Karhunen–Loeve expansion. The fact that a double summation is required in (52) and that the expansion coefficients γ_{kl} require two indices shows the increase in complication that occurs when the orthogonal basis functions are not eigenfunctions.

It is very illuminating to investigate the situation when the expansion (48) turns out to have uncorrelated coefficients. Suppose that in (52) it holds that

$$\gamma_{kl} = \beta_k \delta_{kl} \tag{53}$$

where δ_{kl} is the Kronecker delta. In that case, (52) reduces to

$$c_{XX}(t, s) = \sum_{k=-\infty}^{\infty} \beta_k e^{ik\omega_0(t-s)} \tag{54}$$

that is, $c_{XX}(t, s) = c_{XX}(t - s)$ depends only on the difference $t - s$ and therefore resembles the autocovariance function of a stationary process.

We say "resembles" because stationarity was defined only for processes whose domain of definition is the infinite time interval $-\infty < t < \infty$. In the present case, $\{X(\cdot)\}$ was defined only on $0 \leqslant t \leqslant T$.

Suppose we extend the expansion (48) to values of t outside this interval, still using the same coefficients computed by (49). In that case, of course, the same waveform that occurred on $[0, T]$ repeats itself exactly on successive intervals $[T, 2T]$, $[2T, 3T], \ldots,$ and so on. Thus, the expansion (48) represents a purely periodic process.

Let us suppose we are given a stationary autocovariance function $c_{XX}(\tau)$ defined for $-\infty < \tau < +\infty$. We wish to compare and contrast two possible ways of representing a segment of the associated process $\{X(\cdot)\}$ on a finite time interval. One of the ways will be to construct a periodic process. The function $c_{XX}(\cdot)$ necessarily possesses the symmetry property $c_{XX}(\tau) = c_{XX}^*(-\tau)$. In order to retain this essential property of an autocovariance and also to obtain a function which is periodic with period T, it is necessary to confine attention to the particular interval $[-T/2, T/2]$ rather than $[0, T]$.

There are two procedures open to us, which yield different results.

PROCEDURE 1. Solve the eigenvalue–eigenfunction problem

$$\int_{-T/2}^{T/2} c_{XX}(t-s)\phi_n(s)\,ds = \lambda_n\phi_n(t)$$

This is just (16) adapted to the present situation. The set of functions $\{\phi_n\}$ so obtained yields the expansion

$$c_{XX}(t-s) = \sum_{n=1}^{\infty} \lambda_n\phi_n(t)\phi_n^*(s)$$

which is just (18) adapted to the present case, and the expansion

$$X(t) = \sum_{n=1}^{\infty} \alpha_n\phi_n(t)$$

which is (20).

PROCEDURE 2. Expand the segment of $c_{XX}(\tau)$ falling on the interval $[-T/2, T/2]$ in Fourier series by computing the coefficients

$$\beta_k = \frac{1}{T}\int_{-T/2}^{T/2} c_{XX}(\tau)e^{-ik\omega_0\tau}\,d\tau \tag{55}$$

where, still, $\omega_0 = 2\pi/T$. Use these same coefficients in (54), which gives back c_{XX} with the argument τ replaced by $t - s$.

The point we wish to make now concerns the representation of the function $c_{XX}(t-s)$ in the t, s plane. Equation (54) represents $c_{XX}(t-s)$ on the diagonal strip $-T/2 \leqslant t - s \leqslant T/2$. Outside of this strip, while the originally given function $c_{XX}(t-s)$ behaves as it may, the right-hand side of (54) necessarily repeats over and over on neighboring diagonal strips. In particular (54) does not agree with the originally given c_{XX} on the entire square $-T/2 \leqslant t \leqslant T/2$, $-T/2 \leqslant s \leqslant T/2$.

In contrast, (18) does agree with the original function everywhere on this square. Outside of this square, in general nothing can be said.

The representation of $\{X(\cdot)\}$ associated with Procedure 1 is given by (20). The representation of $\{X(\cdot)\}$ associated with Procedure 2 is given by (48). Both representations do have the properties that the time functions are orthogonal in $L_2[-T/2, T/2]$ and the expansion coefficients are statistically uncorrelated. Therefore, it is legitimate to call both expansions,

"Karhunen–Loeve expansions." This may seem to contradict our earlier statement that this characterization is unique. Actually, there is no contradiction, because we are really talking about two different processes. As the above discussion showed, the two covariance representations do not agree everywhere on the square $-T/2 \leqslant t \leqslant T/2$, $-T/2 \leqslant s \leqslant T/2$, and consequently we have to say that the two processes are different. Nevertheless, at any point t of the interval $[-T/2, T/2]$, both (20) and (48) do converge to the value of the given process in the sense of the discussion pertinent to (27).

See problem 4 at the end of this chapter for a concrete example.

We have taken the segment of $c_{XX}(\tau)$ for $-T/2 \leqslant \tau \leqslant T/2$ and expanded it in a Fourier series. What would happen if we let $T \to \infty$ in an attempt to represent all of c_{XX}? In that case, we make the transition from Fourier series to Fourier integral (Fourier transform), which is familiar from elementary courses (e.g., see Oppenheim and Willsky, 1983, pp. 186–189). In that case, the sums in (48) and (54) become integrals. However, there is a further development, which may be a disaster in some contexts.

On the finite interval, we have

$$E\left\{\int_{-T/2}^{T/2} X^2(t)\,dt\right\} = \int_{-T/2}^{T/2} E\{X^2(t)\}\,dt$$

$$= \int_{-T/2}^{T/2} c_{XX}(0)\,dt = Tc_{XX}(0) \tag{56}$$

where we are assuming c_{XX} is a stationary covariance. Thus, on a finite interval, a segment-stationary process has finite energy content. If we now let $T \to \infty$, we obtain

$$E\left\{\int_{-\infty}^{\infty} X^2(t)\,dt\right\} = \infty \tag{57}$$

which is, of course, mandatory for a stationary process on an infinite interval. On an infinite interval, the process has *infinite energy* but *finite average power*. That is,

$$\lim_{T\to\infty} \frac{1}{2T}\int_{-T}^{T} X^2(t)\,dt < \infty \tag{58}$$

We have dropped the expectation operator in (58) because, if the process is ergodic as well as stationary, the limit in (58) will not be random. It will be the deterministic quantity $c_{XX}(0)$.

Equation (57) indicates that we cannot try to use the Hilbert space $L_2(-\infty, \infty)$ to discuss stationary processes on an infinite interval the way we used $L_2[0, T]$ to discuss (nonstationary) processes on a finite interval. If we try to use the limit on the left-hand side of (58) as the norm of a Hilbert space, it turns out that this is indeed possible. However, the order of infinity of the dimension of the space is no longer countable: it is now uncountable (the precise mathematical terminology would be to say that the Hilbert space is *nonseparable*), which means that there does not exist a countable basis for the space. There simply is *no* representation for functions in that space by discrete sums (no Karhunen–Loeve expansion).

In order to carry this discussion further, we would have to take up the subject of random processes on the frequency axis (the Fourier transform of random processes on the time axis). The rigorous mathematical approach here is through random spectral measures, which is beyond the scope of this book.

Example: Narrow-Band Noise in a Communication System

We now illustrate the results given above by presenting a simplified example of typical design considerations for a digital communications system. Let us consider the transmission of a binary signal by use of coherent phase shift keying over a narrow-band channel with additive noise.

The basic digital signal to be transmitted is a binary sequence of length N, denoted $\{b_m\}$, $m = 1, 2, \ldots, N$. Each character b_m is either 0 or 1. For our example we take the sequence to be deterministic, not random. The sequence is used to modulate a carrier wave which we represent as $A \sin \omega_c t$. An interval of time of total length NT is chosen for transmission, which is divided into N subintervals $(m-1)T \leqslant t \leqslant mT$, $m = 1, 2, \ldots, N$. A typical value of N for some modern communication systems is $N = 1024$. Each subinterval is used to transmit one bit (character) of the sequence $\{b_m\}$. The duration T is chosen so that an exact integral number of cycles of the carrier occurs during each interval: $\omega_c T = 2\pi M$, where M is an integer. In modern sophisticated systems, M can be as small as 1 or 2.

If $b_m = 0$, the carrier is transmitted at its reference phase. If $b_m = 1$, the phase is reversed, that is, shifted by π radians. Therefore, the transmitted signal may be represented as

$$s(t) = A \sin(\omega_c t + \pi b_m)$$

$$(m-1)T \leqslant t \leqslant mT \tag{59}$$

In our model, we will assume that white noise as defined in equation (45) of Chapter 6 is added to $s(t)$ at the input to the channel. The transfer function of the channel will be denoted by $H_c(i\omega)$, with associated impulse response $h_c(t)$. We assume $H_c(i\omega)$ is approximately of the form of the ideal bandpass function given by (54) of Chapter 6, with ω_0 replaced by ω_c, except that H_c meets the Paley–Wiener criterion (50) of Chapter 6, so that it is causal and physically realizable.

Denote the noise at the output of the channel by $V(t)$. We see that $\{V(\cdot)\}$ is a stationary Gaussian process, with mean zero and power spectral density $|H_c(i\omega)|^2$. Define

$$c_{VV}(\tau) = \frac{1}{2\pi}\int_{-\infty}^{\infty} |H_c(i\omega)|^2 e^{-i\omega\tau}\, d\tau \tag{60}$$

Clearly c_{VV} is the autocovariance of $\{V(\cdot)\}$.

Since (59) represents the signal on an individual subinterval, the entire transmitted signal may be written

$$s(t) = A\sum_{m=1}^{N} \{u[t-(m-1)T] - u(t-mT)\}(-1)^{b_m}\sin\omega_c t \tag{61}$$

Here $u(\cdot)$ represents the unit step function. The portion of the channel output due to the signal alone will be denoted by $r(t)$. Evidently

$$r(t) = \int_0^t h_c(t-\tau)s(\tau)\, d\tau \tag{62}$$

Since $h_c(\cdot)$ is causal, $h_c(t) = 0$ for $t < 0$. Let us further assume that the impulse response has decayed sufficiently that it may be considered negligible, for $t > nT$. A typical value for n might be 15.

Let us confine our attention now to a single subinterval $(L-1)T \leqslant t \leqslant LT$, where L is an integer, $n < L < N$. Since the impulse response $h_c(\tau)$ is *not* negligible for $0 \leqslant \tau \leqslant nT$, the present output of the channel will consist of the contribution of the present input bit and the previous n input bits. Making use of the time invariance of the channel and the periodicity of the sine function, we may write the function $r(t)$ in (62), for $(L-1)T \leqslant t \leqslant LT$, in the following way. First, define, for a parameter σ, $0 \leqslant \sigma \leqslant T$, and a set of binary arguments $x_0, x_1, \ldots, x_n$, where each x_k is either 0 or 1, the function

$$\rho(\sigma|x_0, x_1, \ldots, x_n) = A\sum_{q=1}^{q=n}(-1)^{x_q}\int_{\sigma+(q-1)T}^{\sigma+qT} h_c(s)\sin\omega_c(\sigma-s)\, ds$$
$$+(-1)^{x_0}\int_0^{\sigma} h_c(s)\sin\omega_c(\sigma-s)\, ds \tag{63}$$

It follows from (61)–(63), after some manipulation, that for $(L-1)T \leq t \leq LT$,

$$r(t) = \rho(t-(L-1)T|b_L, b_{L-1}, \ldots, b_{L-n}) \tag{64}$$

The signal portion of the channel output, denoted by $r(t)$, will thus consist of a sum of nonoverlapping waveforms, each one defined on an interval of duration T. Alternatively, we may say that (64) provides a piecewise description of $r(t)$, and each different value of the integer L refers to a different segment. Since we have assumed that the b_k's are binary valued, there are 2^{n+1} possible different waveforms. In effect, we have assumed that the channel has a finite memory, where n, the number of intervals of duration T, represents the memory length. The waveforms represented by (64) are sometimes called "chips." The waveform actually received at the output of the channel will be denoted by $X(\cdot)$. It is the sum of the signal and noise:

$$X(t) = r(t) + V(t) \tag{65}$$

We now describe the structure of a suboptimal receiver for demodulating the received waveform and estimating the input bit string. The idea is to base the scheme on the use of the Fourier coefficients ζ_k in (48). Since there are infinitely many coefficients, and since any practical scheme will be limited to using a finite number of them, let us first truncate (48) in some way that yields a suitable approximation. The two relationships $\omega_0 = 2\pi/T$ and $\omega_c T = 2\pi M$ together imply

$$\omega_c = M\omega_0 \tag{66}$$

The most important terms in (48) can be expected to be those centered about the carrier frequency ω_c, recalling our assumption concerning the bandpass nature of the channel transfer function $H_c(i\omega)$.

Therefore, the most important terms in (48) will occur for values of the index k near either $-M$ or $+M$. Adopt the integer B as the number of terms on either side of the carrier that should be kept in order to have a suitable approximation. A reasonable value of B might be the integer next larger than Ω/ω_0, where Ω is the one-sided bandwidth of the channel. The resulting approximation to (48) may be written

$$X(t) \sim \sum_{k=-(M+B)}^{k=-(M-B)} \zeta_k e^{ik\omega_0 t} + \sum_{k=M-B}^{k=M+B} \zeta_k e^{ik\omega_0 t} \tag{67}$$

The first summation contains the terms for negative values of the index k, while the second summation contains the terms for positive values. Because $X(t)$ is an actual, physically measurable signal, it must be purely real-valued, so that the coefficients ζ_k in the first sum are merely the complex conjugates of those in the second. From (65), we see that the random component of X is the narrow-band noise V, so here we are applying the technique of the previous section to this noise.

The set of coefficients $\{\zeta_{M-B}, \ldots, \zeta_{M+B}\}$, which is a finite set of numbers, may be considered as a complex-valued finite-dimensional vector in some "signal space" which approximately represents the received waveform. The task now is to devise a way to use them to estimate the present input bit b_L. The point of using the Fourier expansion in the example is just to achieve this vector representation.

We may assume that at time $t = (L-1)T$, we already have our estimates $\hat{b}_{L-n}, \hat{b}_{L-n+1}, \ldots, \hat{b}_{L-1}$ of the input bits over the n previous intervals. Since the mean of V is zero, we have from (64) and (65) that for $(L-1)T \leqslant t \leqslant LT$,

$$\mu_X(t) = E\{X(t)\} = \rho(t - (L-1)T | b_L, b_{L-1}, \ldots, b_{L-n}) \tag{68}$$

Motivated by the coefficients $\bar{\zeta}_k$ defined in (50), let us now use the function $\rho(\sigma | x_0, x_1, \ldots, x_n)$ defined in (63) to define the following quantities

$$\bar{\zeta}_k(x_0, x_1, \ldots, x_n) = \frac{1}{T}\int_0^T \rho(\sigma | x_0, \ldots, x_n) e^{-ik\omega_0\sigma}\, d\sigma \tag{69}$$

for $M - B \leqslant k \leqslant M + B$, k integer. Each quantity $\bar{\zeta}_k$ is a function of the set of binary variables $x_0, x_1, \ldots, x_n$. These quantities are completely defined in terms of known properties of the channel and the transmitted signal, so they may be precomputed before the communication system ever starts operating.

When the receiver is in operation, let it compute in real time, anew on each subinterval $(L-1)T \leqslant t \leqslant LT$, the complex coefficients

$$\hat{\zeta}_k = \frac{1}{T}\int_{(L-1)T}^{LT} X(t) e^{-ik\omega_0 t}\, dt \tag{70}$$

for $M - B \leqslant k \leqslant M + B$, k integer, using the actual received waveform $X(\cdot)$. The idea proposed here is to compare the coefficients $\hat{\zeta}_k$ from (70) to the $\bar{\zeta}_k$ in (69) in such a way that the current bit b_L can be estimated. To this end define the two "distances"

$$d(0) = \sum_{k=M-B}^{k=M+B} |\hat{\zeta}_k - \bar{\zeta}_k(0, \hat{b}_{L-1}, \ldots, \hat{b}_{L-n})|^2 \tag{71}$$

$$d(1) = \sum_{k=M-B}^{k=M+B} |\hat{\zeta}_k - \bar{\zeta}_k(1, \hat{b}_{L-1}, \ldots, \hat{b}_{L-n})|^2 \tag{72}$$

In (71) and (72), we are inserting our estimates $\hat{b}_{L-1}, \ldots, \hat{b}_{L-n}$ of the previous bits into the functions $\bar{\zeta}_k$. We use both possible values, 0 and 1, of the present input bit b_L to obtain *two* possible values of each coefficient $\bar{\zeta}_k$. The two distances $d(0)$ and $d(1)$ are indications of the respective agreements of our actual waveform coefficients $\hat{\zeta}_K$ with our two possible values of $\bar{\zeta}_k$. The detection now is simple:

$$\text{If } d(0) > d(1), \text{ estimate } \hat{b}_L = 1$$

$$\text{If } d(0) < d(1), \text{ estimate } \hat{b}_L = 0$$

The expansion (67) on which this scheme is based is actually Procedure 2 of the previous section applied to the received waveform. An alternative approach would be to use Procedure 1. This discussion was intended to illustrate the application of that theory, but not to present detailed principles for receiver design.

A block diagram of this receiver is shown in Figure 8.1 Because of the linearity of the channel and the operation (70), the random variables $\hat{\zeta}_k$ will have Gaussian distributions if the noise is Gaussian. Thus, with more persistence one could pursue this investigation into computing the probability of error (decoding a 1 when a 0 was sent and vice versa) as a function of signal to noise ratio, which is controlled by the parameter A in (59). This is

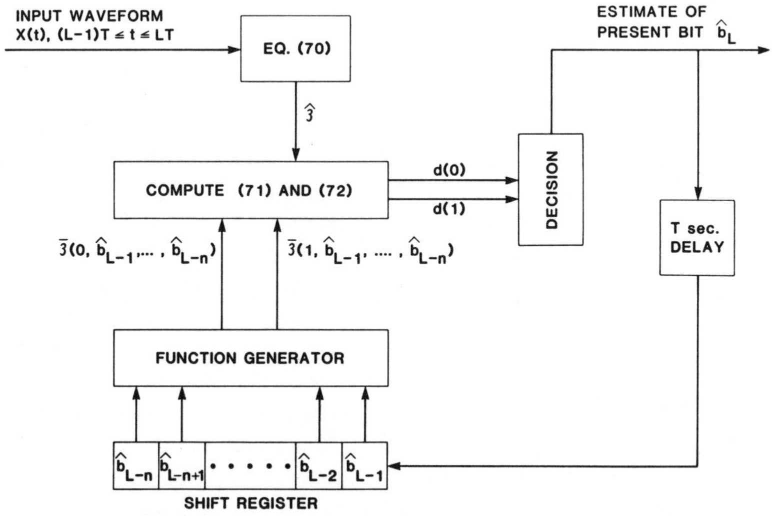

FIGURE 8.1. Block diagram of sub-optimal receiver.

the type of analysis that has to be carried out in the design of an actual system, but it goes beyond the intended scope of the present discussion.

The Uncertainty Principle

The topics discussed so far put us in a position to consider a result that is known as the Uncertainty Principle. A qualitative statement of it is that there are limits to the extent to which it is possible simultaneously to confine the energy content of a signal in both time and frequency. A special case is the statement that no continuous time signal can vanish identically everywhere outside a finite time interval and simultaneously possess a spectrum that vanishes identically everywhere outside of a finite frequency interval.

In this book we content ourselves with examining some specific cases of this principle, stating some of the precise mathematical formulations of the principle and citing references where complete proofs may be found. The principle itself is actually an inherent consequence of the Fourier transform, and applies only to deterministic signals. The reason for including it in this book is that typical applications of the stochastic process theory currently being discussed also require some familiarity with signal analysis and signal design for deterministic signals. Therefore, it seems appropriate to include a few items from the latter topic in this course.

We use the symbol $u(t)$ to denote the unit step, and $p(t)$ to denote a rectangular pulse:

$$u(t) = \begin{cases} 0, & t < 0 \\ 1, & t \geqslant 0 \end{cases}$$
$$p(t) = u(t+1) - u(t-1) \tag{73}$$

In Table 8.1 we list certain signals along with their Fourier transforms.

The first case in Table 8.1 is $f(t) = \cos \omega_0 t$, a continuous single frequency carrier wave of infinite duration. Its transform, $F(i\omega)$, consists of two δ-functions located at $+\omega_0$ and $-\omega_0$, respectively. Thus, all of the energy is concentrated at two points in the frequency domain, but it is totally spread out in the time domain. In case 2, the cosine wave is only of semi-infinite duration, being turned on at $t = 0$. The transform is still concentrated at $\pm\omega_0$, but now $F(i\omega)$ has developed "skirts" or "tails," so there is a continuous distribution of energy in the frequency domain in addition to the concentration at $\pm\omega_0$. In case 3, we have a rectangular pulse of carrier wave, of finite duration. Its transform is now bounded at all frequencies, and the energy distribution is still more spread out. Case 4 is the extreme case of concentrating energy at one point in the time domain. Its transform $F(i\omega)$ now has unit magnitude at all frequencies.

TABLE 8.1. Some Selected Fourier Transform Pairs

$$F(i\omega) = \int_{-\infty}^{\infty} f(t) e^{-i\omega t}\, dt$$

Case	$f(t)$	$F(i\omega)$
1	$\cos \omega_0 t$	$\pi[\delta(\omega - \omega_0) + \delta(\omega + \omega_0)]$
2	$u(t) \cos \omega_0 t$	$\frac{\pi}{2}[\delta(\omega - \omega_0) + \delta(\omega + \omega_0)] + \frac{i\omega}{\omega_0^2 - \omega^2}$
3	$p(t) \cos \omega_0 t$	$\frac{\sin(\omega + \omega_0)}{(\omega + \omega_0)} + \frac{\sin(\omega - \omega_0)}{(\omega - \omega_0)}$
4	$\delta(t - t_0)$	$e^{-i\omega t_0}$
5	$e^{-\|t\|} \cos \omega_0 t$	$\frac{1}{1 + (\omega - \omega_0)^2} + \frac{1}{1 + (\omega + \omega_0^2)}$
6	$\frac{1}{\sqrt{\pi}} e^{-t^2} \cos \omega_0 t$	$\frac{e^{-(\omega - \omega_0)^2/4} + e^{-(\omega + \omega_0)^2/4}}{2}$

These four examples should serve as a qualitative review of a phenomenon of which the student is probably already aware: attempting to concentrate energy in one domain tends to spread it out in the other. The last two examples are a cosine carrier wave modulated respectively with an exponential envelope and a Gaussian envelope. We will use these to study this phenomenon more quantitatively.

Actually, since the phenomenon is an attribute of the envelope rather than the carrier, we will study the low-pass version of the bandpass cases 5 and 6. For positive, real values of the parameters T and Ω, let us define the quantities

$$\alpha^2 = \frac{\int_{-T/2}^{T/2} |f(t)|^2\, dt}{\int_{-\infty}^{\infty} |f(t)|^2\, dt} \tag{74}$$

$$\beta^2 = \frac{\int_{-\Omega}^{\Omega} |F(i\omega)|^2\, d\omega}{\int_{-\infty}^{\infty} |F(i\omega)|^2\, d\omega} \tag{75}$$

TABLE 8.2. Fractional Energy Content of a Waveform and Its Spectrum

$f(t) = e^{-|t|}$ $\qquad$ $F(i\omega) = \dfrac{2}{\omega^2 + 1}$

$\alpha^2 = 1 - e^{-T}$ $\qquad$ $\beta^2 = \dfrac{2}{\pi}\left(\tan^{-1}\Omega + \dfrac{\Omega}{\Omega^2 + 1}\right)$

T	α^2	Ω	β^2
1	0.63212	1	0.81831
2	0.86466	2	0.95948
3	0.95021	3	0.98615

For the low-pass version of case 5, we have $f(t) = e^{-|t|}$, $F(i\omega) = 2/(\omega^2 + 1)$, and by direct calculation one finds

$$\alpha^2 = 1 - e^{-T} \tag{76}$$

$$\beta^2 = \frac{2}{\pi}\left[\tan^{-1}\Omega + \frac{\Omega}{\Omega^2 + 1}\right] \tag{77}$$

We will be particularly interested in the product ΩT as a measure of the degree of concentration of energy in both domains simultaneously. From (75), (76), α^2 is the fraction of the total energy of the waveform which lies in the interval $[-T/2, T/2]$, while β^2 is the fraction of total energy in the spectrum which lies in the interval $[-\Omega, \Omega]$. Table 8.2 summarizes the results for this example.

The nature of the calculations is such that it is necessary to pick values for T and Ω, and then evaluate the resulting values of α^2 and β^2. However, in applying these results to system design, the natural approach would be the opposite. For example, suppose we wanted to insure that 95% of the energy in the waveform is contained between $-T/2$ and $T/2$. According to this table, we must choose $T \geqslant 3$. If simultaneously we wanted to insure that 95% of the energy in the spectrum lies between $-\Omega$ and Ω, we must choose $\Omega \geqslant 2$.

In this case, the waveform $f(t)$ is not totally concentrated on any finite interval of time, nor is the spectrum $F(i\omega)$ totally concentrated on any finite interval of frequency. The best we can do is, as above, determine those finite intervals on which a specified fraction of the total energy is concentrated. As this fraction is pushed upward toward 1.0, the length of the corresponding interval goes to infinity.

TABLE 8.3. Fractional Energy Content of a Waveform and Its Spectrum

$f(t) = \frac{1}{\sqrt{\pi}} e^{-t^2}$ $\quad F(i\omega) = e^{-\omega^2/4}$

Define $\Phi(u) = \int_0^u \frac{1}{\sqrt{2\pi}} e^{-x^2/2}\, dx$. Then

$\alpha^2 = 2\Phi(T)$		$\beta^2 = 2\Phi(\Omega)$	
T	α^2	Ω	β^2
1	0.6826	1	0.6826
2	0.9546	2	0.9546
3	0.9974	3	0.9974

In Table 8.3 we summarize the results for the low-pass version of case 6 of Table 8.1. Comparison of the results shows that the Gaussian envelope concentrates a larger fraction of energy into a specified interval, for any given time-bandwidth product. Conversely, if the fraction of energy to be concentrated, that is, the values of α^2 and β^2, are specified, then a smaller value of ΩT can be achieved for the Gaussian envelope than for the exponential.

The question arises as to what is the best that can be done in this regard. This question of "best" is conveniently stated mathematically in terms of optimization in Hilbert space. Consider $L_2(-\infty, \infty)$ as a repository of waveforms. Clearly, the values of α^2 and β^2 in (74) and (75) are unaffected by multiplying a given function f by a constant scale factor. Therefore, we may confine attention to the set of functions having $\|f\| = 1$. Within this set, one way of formulating the problem is to consider the subset of functions all achieving at least some minimum value of α. Taking this subset for some given value of $\alpha < 1$, we then seek that particular function which achieves the highest value of β. Throughout this, we are assuming the values of T and Ω to be fixed.

The next point is that by using the Parseval equality, one has

$$\int_{-\infty}^{\infty} |f(t)|^2\, dt = \frac{1}{2\pi} \int_{-\infty}^{\infty} |F(i\omega)|^2\, d\omega \tag{78}$$

and by using properties of Fourier transforms, the identity can be proven that

$$\int_{-\Omega}^{\Omega} |F(i\omega)|^2\, d\omega = \frac{1}{\pi} \int_{-\infty}^{\infty} \int_{-\infty}^{\infty} \frac{\sin \Omega(t-s)}{t-s} f(t) f(s)\, dt\, ds \tag{79}$$

Therefore, the problem of simultaneously concentrating energy in both time and frequency can be studied in terms of the following quadratic optimization problem in Hilbert space:

Choose f in $L_2(-\infty, \infty)$ to maximize the quantity on the right-hand side of (79), subject to the inequality constraint

$$\int_{-T/2}^{T/2} |f(t)|^2 \, dt \geqslant \alpha^2 \tag{80}$$

and the equality constraint

$$\int_{-\infty}^{\infty} |f(t)|^2 \, dt = 1 \tag{81}$$

The problem is addressed this way in the paper by Landau and Pollack (1961). In the examples given in Tables 8.2 and 8.3, the function f was specified. If f is specified, then time and frequency are "decoupled" in the sense that α^2 is a function of T only and β^2 is a function of Ω only. However, in the optimization problem we are presently considering, Ω, T, and α are specified. The object is then to search through $L_2(-\infty, \infty)$ to find the function f that achieves the largest β^2, which is what (79) is when the constraints (80) and (81) are met. By means of this approach, Landau and Pollack were able to show the following result. There exists a function $\Phi(\alpha, \beta)$, such that the product ΩT obeys the following inequality:

$$\Omega T \geqslant \Phi(\alpha, \beta) \tag{82}$$

The function Φ is such that when both $\alpha \to 1$ and $\beta \to 1$ simultaneously, then $\Phi \to \infty$. This is the special case of the uncertainty principle to which we alluded at the beginning of this section: it is impossible for a function simultaneously to have finite duration in time and finite bandwidth in frequency.

We conclude this chapter by stating the results of Landau and Pollack slightly more explicitly. They showed that the solution to the quadratic optimization problem, maximize (79) subject to (80) and (81), hinges on finding the largest eigenvalue λ_0 and the associated eigenfunction f_0 of the following integral operator eigenvalue problem:

$$\frac{1}{\pi} \int_{-T/2}^{T/2} \frac{\sin \Omega(t-s)}{t-s} f(s) \, ds = \lambda f(t) \tag{83}$$

It turns out that this largest eigenvalue λ_0 is a continuous monotonically strictly increasing function of the product ΩT. For $\Omega T = 0$, $\lambda_0 = 0$, and for

$\Omega T \to \infty$, $\lambda_0 \to 1$. Therefore, there exists an inverse function

$$\Omega T = \phi(\lambda_0) \tag{84}$$

again continuous and strictly monotonically increasing, with $\phi(0) = 0$ and $\phi(\lambda_0) \to \infty$ as $\lambda_0 \to 1$.

By use of geometric considerations in Hilbert space, Landau and Pollack obtained the inequality

$$\cos^{-1}\alpha + \cos^{-1}\beta \geqslant \cos^{-1}\sqrt{\lambda_0} \tag{85}$$

Now use the trigonometric identities, valid in the first quadrant

$$\begin{aligned} \sin(\cos^{-1}x) &= \sqrt{1 - x^2} \\ \sin(\cos^{-1}\alpha + \cos^{-1}\beta) &= \beta\sqrt{1-\alpha^2} + \alpha\sqrt{1-\beta^2} \end{aligned} \tag{86}$$

By (86), (85) is equivalent to

$$\lambda_0 \geqslant 1 + 2\alpha^2\beta^2 - \alpha^2 - \beta^2 - 2\alpha\beta\sqrt{(1-\alpha^2)(1-\beta^2)} \tag{87}$$

provided that $\alpha^2 + \beta^2 \geqslant 1$.

Since the function ϕ in (84) is monotone increasing, if we take that function of both sides of (87), the inequality still holds in the same direction:

$$\phi(\lambda_0) \geqslant \phi\left(1 + 2\alpha^2\beta^2 - \alpha^2 - \beta^2 - 2\alpha\beta\sqrt{(1-\alpha^2)(1-\beta^2)}\right)$$
$$\alpha^2 + \beta^2 \geqslant 1 \tag{88}$$

Combining (84) with (88) we have the promised inequality (82). This is the time–frequency uncertainty principle. For $\alpha^2 + \beta^2 = 1$, one finds $\Omega T \geqslant 0$, which is necessarily true. Thus, for values of α^2 and β^2 such that $\alpha^2 + \beta^2 < 1$, there is no constraint on ΩT, and the inequality here is inapplicable.

It is important to note that its proof involved the eigenvalue problem (83), which is similar in form to the problem (16) that we discussed at the beginning of this chapter. Thus, there is the common thread of the idea of representing a function or a waveform by means of a spectral representation utilizing the eigenvalue problem in Hilbert space, which runs through this entire chapter and provides a unifying theme.

Problems

1. Given $f(t)$ in $L_2[0, \pi]$, let $g(t)$ be that linear combination of $\sin t, \sin 2t, \sin 3t$ which minimizes

$$\int_0^{\pi} [f(t) - g(t)]^2 \, dt = \|f - g\|^2$$

Show that $g(t)$ is the projection of $f(t)$ on the subspace spanned by $\sin t, \sin 2t, \sin 3t$, that is, show that the error $f(t) - g(t)$ is orthogonal to any linear combination of $\sin t, \sin 2t, \sin 3t$.

2. (Courant–Hilbert). Given the kernel

$$K(s, t) = \tfrac{1}{2} \log \left| \frac{\sin\left(\frac{s+t}{2}\right)}{\sin\left(\frac{s-t}{2}\right)} \right|$$

Show that the eigenvector/eigenvalue problem

$$\int_0^{\pi} K(s, t) \phi_n(t) \, dt = \lambda_n \phi_n(s)$$

has eigenvalues $\lambda_n = 2n/\pi$ and eigenfunctions $\phi_n(t) = \sin nt$, $n = 1, 2, \ldots$.

3. (Courant–Hilbert). Given the kernel

$$K(s, t) = \frac{1}{2\pi} \frac{1 - h^2}{1 - 2h \cos(s - t) + h^2}$$

with $|h| < 1$. Show that the eigenvalue/eigenvector problem

$$\int_0^{2\pi} K(s, t) \phi_n(t) \, dt = \lambda_n \phi_n(s)$$

has two sets of eigenfunctions $\{\psi_n\}$ and $\{\phi_n\}$, where

$$\phi_n(t) = \cos nt, \qquad n = 0, 1, 2, \ldots$$

with corresponding eigenvalues

$$\lambda_n = h^{-n} \quad \text{and} \quad \psi_n(t) = \sin nt, \qquad n = 1, 2, \ldots$$

with corresponding eigenvalues

$$\lambda_n = h^{-n}$$

4. Let $\{X(\cdot)\}$ be the stationary Gaussian process with mean zero and autocovariance $c_{XX}(\tau) = e^{-|\tau|}$.

PROCEDURE 1. Solve the eigenvalue–eigenfunction problem

$$\int_{-T/2}^{T/2} e^{-|t-s|}\phi_n(s)\,ds = \lambda_n\phi_n(t)$$

This can be done by breaking up the integral in order to eliminate the absolute value sign, and then differentiating the resulting expression with respect to t to obtain a differential equation for ϕ_n, with appropriate boundary conditions. If you do this correctly, you will obtain the result that, although the ϕ_n are trigonometric functions, the λ_n are not harmonically related.

PROCEDURE 2. Compute the Fourier coefficients

$$\beta_k = \frac{1}{T}\int_{-T/2}^{T/2} e^{-|\tau|}e^{-ik\omega_0\tau}\,d\tau, \qquad \omega_0 = \frac{2\pi}{T}$$

This can be simplified by using

$$e^{-ik\omega_0\tau} = \cos k\omega_0\tau - i\sin k\omega_0\tau$$

and noting that because of symmetry, only the cosine term contributes. Therefore, by symmetry,

$$\beta_k = \frac{2}{T}\int_0^{T/2} e^{-\tau}\cos k\omega_0\tau\,d\tau$$

Write out the expressions for $e^{-|t-s|}$ corresponding to (18) and (54), respectively, and study their behavior in the t, s plane.

5. Let $f(t)$ be the triangular pulse

$$f(t) = \begin{cases} 0, & t < -1 \\ t+1, & -1 \leqslant t < 0 \\ 1-t, & 0 \leqslant t < 1 \\ 0, & t \geqslant 1 \end{cases}$$

For this $f(t)$, find its Fourier transform $F(i\omega)$. Compute α^2 for $T = 1, 2, 3$, and β^2 for $\Omega = 1, 2, 3$ as in Tables 8.2 and 8.3.

Chapter 9

Linear Systems in Conjunction with Memoryless Nonlinear Devices

Introduction

In this chapter, we depart from the focus of the rest of the book on strictly linear systems and discuss the passage of stochastic processes through systems containing certain simple types of nonlinearities. Our intention here is not to provide a thorough review of all the literature on this subject, but merely to give a partial survey of some of the principal analysis techniques that have been popular or have been found effective.

Nonlinearities are deliberately introduced into communication systems in two important applications: modulation and detection. In our first example, we discuss a device that has been used as a model for a noncoherent demodulator of *AM* signals, the square-law detector.

The Square-Law Detector

The discussion will be facilitated by beginning with a brief elementary review of the operation of conventional amplitude modulation and demodulation for deterministic signals, in a system using a square-law device as a detector. The word "detector" in this context is simply traditional usage and is synonymous with "demodulator."

The system under consideration is illustrated in Figure 9.1. The modulating signal is indicated as $m(t)$, and the carrier wave is designated $\cos \omega_c t$. In order to avoid so-called "overmodulation," the amplitude of the input must be restricted by the condition $|m(t)| < 1$. A constant bias of unity is added

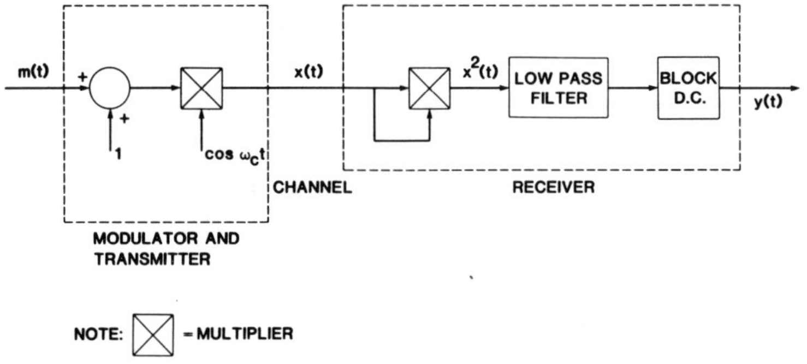

FIGURE 9.1. An AM transmitter and receiver with square-law detector.

to the signal $m(t)$, and the sum multiplies the carrier wave to yield the transmitted signal $x(t)$. Assuming an ideal channel, the received signal is also $x(t)$.

As shown in Figure 9.1, the simplified receiver consists of a square-law device followed by a low-pass filter. From the diagram we have

$$x(t) = [1 + m(t)]\cos \omega_c t \tag{1}$$

whence, using the identity $\cos^2 z = (1 + \cos 2z)/2$,

$$x^2(t) = [1 + 2m(t) + m^2(t)]\left(\frac{1 + \cos 2\omega_c t}{2}\right) \tag{2}$$

Assuming that $m(t)$ is slowly varying in comparison with $\cos \omega_c t$, the low-pass filter is designed to pass all frequencies from zero up to the highest frequency required to reproduce $m(t)$ faithfully. It can be assumed that the product $(1 + m)^2 \cos \omega_c t$ is excluded. The output of the low-pass filter is therefore proportional to $1 + 2m(t) + m^2(t)$. The constant bias of unity is now removed by a blocking capacitor. In early applications, $m(t)$ was merely an audio signal, so it sufficed to preserve frequencies from 30 hertz or so on up to 5000 hertz, depending on the fidelity being demanded. The final output, as shown in the diagram, is then

$$y(t) = m(t) + \tfrac{1}{2}m^2(t) \tag{3}$$

The term $\frac{1}{2}m^2(t)$ is the so-called quadratic distortion, introduced by the square-law device. It might be acceptable to allow this amount of distortion in speech channels, but it would not be acceptable in high-fidelity music applications. There, a detector having an absolute value $|x|$ characteristic would be employed, but the analysis of that becomes more intricate. The square-law device allows a relatively simple analysis for the purpose of developing a qualitative conceptualization of the operation of the communications system.

Let us now replace the deterministic signals with stochastic processes, in order to study our first example of stochastic processes in a system containing a nonlinearity. It is desirable to do this in such a way that we can employ the methods presented for analysis of stationary second-order processes in Chapter 6. Therefore, it is worthwhile to take a little time to select the most appropriate model and set of assumptions before we start analyzing.

If we simply try to replace the deterministic signal $m(t)$ in the previous example by a stochastic process $M(t)$, we run into trouble, because the process $[1 + M(t)]\cos \omega_c t$ is *not stationary*. Observe:

$$E\{[1 + M(t)][1 + M(s)]\cos \omega_c t \cos \omega_c s\}$$

$$= \{1 + E[M(t)] + E[M(s)] + E[M(t)M(s)]\}\cos \omega_c t \cos \omega_c s \quad (4)$$

The difficulty is that $\cos \omega_c t \cos \omega_c s$ cannot be written as a function of $t - s$ only. Therefore, even if $\{M(\cdot)\}$ itself is stationary, there is no power spectral density corresponding to (4). Since it is mainly the properties of the spectrum that we want to discuss, it will be necessary to proceed more circumspectly.

It turns out that the most fruitful way to proceed is simply to assume that $\{X(\cdot)\}$ is a stationary process with a power spectrum of the so-called bandpass form, which can be related to a corresponding low-pass form of power spectrum belonging to another stationary process $\{M(\cdot)\}$. In formulating our assumptions in this way, we sidestep altogether the question of modeling the modulator as we did in Figure 9.1. Rather, we pick up the analysis at the receiver and proceed from there.

Because it is a theme which will recur throughout this chapter, it is appropriate at this point to digress for a moment in order to discuss the meaning of the terms "low pass" and "band pass." The terms arose in practice and have never been made precise in a way that is satisfactory to all users. Roughly, "band pass" means that the spectrum of the signal is mainly concentrated about a frequency ω_c which is removed from the origin by an amount greater than the width of the concentration. Such signals typically arise by modulating a carrier wave with some other input. The input prior to modulation is often referred to as a *low-pass* or *baseband signal*.

In mathematics, if a function $F(\omega)$ vanishes identically outside of some interval $a \leq \omega \leq b$, denoted $[a, b]$, it is customary to say that $[a, b]$ is the *support* of the function F. Because of difficulties arising from the Paley–Wiener criterion, the Uncertainty Principle, and their implications for the spectra of physical systems, it is too restrictive to postulate systems having spectra that actually vanish identically outside some such bounded interval. For this book, we will say an interval $[a, b]$ is the *effective support* of a function F, if the magnitude of F is negligible outside of $[a, b]$. In that case, we can proceed for the purposes of the argument at hand as if F in fact vanished identically outside of $[a, b]$, provided we realize that any conclusions reached in this fashion may be only approximately valid.

This line of thought is roughly what is meant by communications engineers when they speak of the *narrow-band approximation*. Because of the sheer complexity of communications and radar systems, historically it has been found indispensable to be able to pursue such a line of reasoning, even though there are difficulties in making it mathematically precise.

At the end of this chapter we will present some more techniques that are useful in studying and analyzing the processes of modulation and demodulation. To discuss those now would take us further afield than we wish to go at the moment. All we want to do for the present time is to introduce the study of systems containing a nonlinearity, and we happen to have chosen a demodulator as an illustration.

Having made these observations, it now seems safe to proceed.

In Figure 9.2 we repeat the diagram of the square-law detector. A band-pass process $X(t)$ is passed through a square-law nonlinearity, followed by a low-pass filter, to yield an output process $Z(t)$. The intention behind this is that the spectrum of the $Z(\cdot)$ process should resemble the low-pass version of the spectrum of the $X(\cdot)$ process. Let us consider the details of the analysis.

In Figure 9.3 we have sketched a typical power spectrum $\phi_{XX}(\omega)$ of the $X(\cdot)$ process. It shows that the spectrum is negligible outside of bands of width 2Ω centered at $\pm\omega_c$. Specifically, we are assuming that the effective

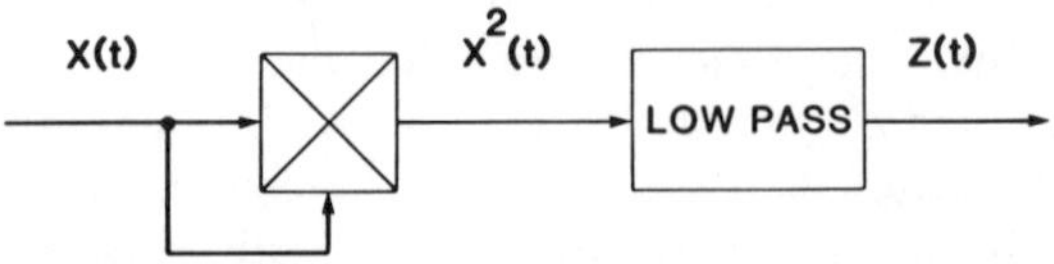

FIGURE 9.2. Close-up of square-law demodulator.

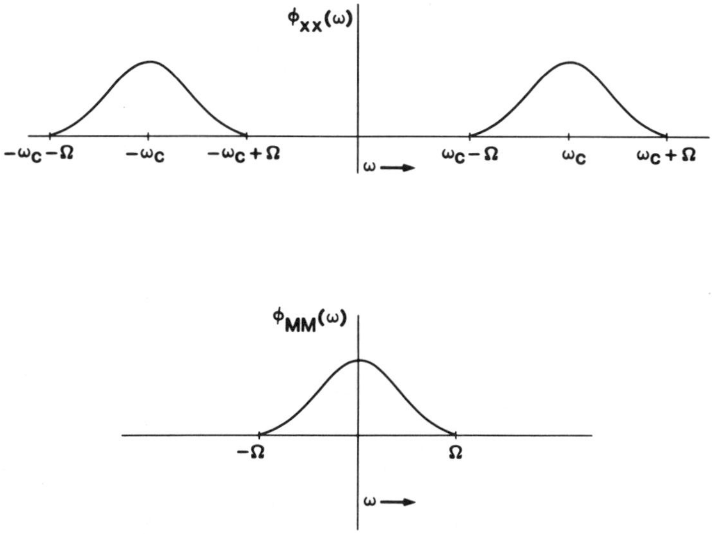

FIGURE 9.3. Plots of power spectral densities ϕ_{XX} and ϕ_{MM}.

support of $\phi_{XX}(\omega)$ is the set

$$[-\omega_c - \Omega, -\omega_c + \Omega] \cup [\omega_c - \Omega, \omega_c + \Omega]$$

A sketch of $\phi_{MM}(\omega)$ also appears in Figure 9.3, representing a typical low-pass spectrum. Here we are assuming the effective support to be the set $[-\Omega, \Omega]$.

The assumption $\Omega \ll \omega_c$ is what we mean by the *narrow-band* assumption concerning ϕ_{XX}. Under these assumptions, it is possible to write

$$\phi_{XX}(\omega) = \phi_{MM}(\omega - \omega_c) + \phi_{MM}(\omega + \omega_c) \tag{5}$$

Equation (5) may be written as a convolution:

$$\phi_{XX}(\omega) = \int_{-\infty}^{\infty} \phi_{MM}(\omega - u)[\delta(u - \omega_c) + \delta(u + \omega_c)]\, du \tag{6}$$

Let ϕ_{XX} and ϕ_{MM}, respectively, be the Fourier transforms of autocovariance functions $c_{XX}(\tau)$ and $c_{MM}(\tau)$. Recalling the convolution theorem, the

Fourier transform of (6) is

$$c_{XX}(\tau) = \frac{1}{\pi} c_{MM}(\tau) \cos \omega_c \tau \tag{7}$$

For convenience, introduce the notation

$$Y(t) = X^2(t) \tag{8}$$

We are interested in the autocovariance

$$c_{YY}(\tau) = E\{[Y(t+\tau) - \mu_Y][Y(t) - \mu_y]\} \tag{9}$$

Since $\{X(\cdot)\}$ is stationary, so is $\{Y(\cdot)\}$, so μ_Y must be constant. Its value is

$$\mu_Y = E\{Y(t)\} = E\{X^2(t)\} = c_{XX}(0) + \mu_X^2 \tag{10}$$

Since $\{X(\cdot)\}$ is stationary, if μ_X is not zero, it must be constant. However, because we have assumed $\{X(\cdot)\}$ to be a band-pass process, it is inconsistent to assume there is a "d.c." component present. Therefore, assume $\mu_X = 0$. Equation (9) can be written

$$c_{YY}(\tau) = E\{Y(t+\tau)Y(t)\} - \mu_Y^2 \tag{11}$$

From (8),

$$E\{Y(t+\tau)Y(t)\} = E\{X^2(t+\tau)X^2(t)\} \tag{12}$$

Consider the joint probability density of the two r.v.'s $X(t+\tau)$ and $X(t)$. Suppose $\{X(\cdot)\}$ is a Gaussian process, with mean zero and autocovariance c_{XX}. Define $f(x_1, x_2)$ by

$$f(x_1, x_2)\, dx_1\, dx_2$$
$$= P\{x_1 \le X(t+\tau) < x_1 + dx_1, x_2 \le X(t) < x_2 + dx_2\}$$

Then

$$f(x_1, x_2) = \frac{|\mathbf{C}|^{-1/2}}{2\pi} \exp\left\{-\tfrac{1}{2}[x_1 \quad x_2]\mathbf{C}^{-1}\begin{bmatrix} x_1 \\ x_2 \end{bmatrix}\right\} \tag{13}$$

where the covariance matrix $\mathbf{C}$ is given by

$$\mathbf{C} = \begin{bmatrix} c_{XX}(0) & c_{XX}(\tau) \\ c_{XX}(\tau) & c_{XX}(0) \end{bmatrix} \tag{14}$$

As discussed in Chapter 1, the associated characteristic function $M(u_1, u_2)$ is

$$M(u_1, u_2) = \int_{-\infty}^{\infty}\int_{-\infty}^{\infty} f(x_1, x_2)e^{-iu_1x_1}e^{-iu_2x_2}\,dx_1\,dx_2$$

In this case, it turns out explicitly

$$M(u_1, u_2) = \exp\left\{-\tfrac{1}{2}\left[u_1^2c_{XX}(0) + 2u_1u_2c_{XX}(\tau) + u_2^2c_{xx(0)}\right]\right\} \quad (15)$$

By the properties of the characteristic function

$$E\{X^2(t+\tau)X^2(t)\} = \left.\frac{\partial^4 M(u_1, u_2)}{\partial u_1^2\,\partial u_2^2}\right|_{\substack{u_1=0\\u_2=0}} \quad (16)$$

Evaluation of the right-hand side of (16), using (15), is left as a homework problem. The result is

$$E\{X^2(t+\tau)X^2(t)\} = 2c_{XX}^2(\tau) + c_{XX}^2(0) \quad (17)$$

From (10)–(12), therefore

$$c_{YY}(\tau) = 2c_{XX}^2(\tau) \quad (18)$$

Using the identity $\cos^2\theta = (1 + \cos 2\theta)/2$, and substituting (7) into (18), we have

$$c_{YY}(\tau) = \pi^{-2}c_{MM}^2(\tau)[1 + \cos 2\omega_c\tau] \quad (19)$$

The Fourier transform of $c_{MM}(\tau)$ is, by definition, the power spectrum $\phi_{MM}(\omega)$. By the convolution theorem, the Fourier transform of $c_{MM}^2(\tau)$ is ϕ_{MM} convolved with itself, which we denote by $\psi(\omega)$:

$$\psi(\omega) = \mathcal{F}\{c_{MM}^2\} = \int_{-\infty}^{\infty} \phi_{MM}(\omega - u)\phi_{MM}(u)\,du \quad (20)$$

The Fourier transform of $\pi^{-2}[1 + \cos 2\omega_c\tau]$ is $(1/\pi)[2\delta(\omega) + \delta(\omega - 2\omega_c) + \delta(\omega + 2\omega_c)]$. The Fourier transform of $c_{YY}(\tau)$, which is the power spectrum $\phi_{YY}(\omega)$, is therefore the convolution of this combination of δ-functions with $\psi(\omega)$. Explicitly, therefore,

$$\phi_{YY} = \frac{1}{\pi}[2\psi(\omega) + \psi(\omega - 2\omega_c) + \psi(\omega + 2\omega_c)] \quad (21)$$

Let the low-pass filter in Figure 9.2 have transfer function $H(i\omega)$. Then

$$\phi_{ZZ}(\omega) = |H(i\omega)|^2\phi_{YY}(\omega) \tag{22}$$

[Recall equation (8).] If we assume that $|H(i\omega)|$ is negligible for $|\omega| > \Omega$ and that $\Omega \ll \omega_c$, then we may replace (22) by

$$\phi_{ZZ}(\omega) = \frac{2}{\pi}|H(i\omega)|^2\psi(\omega) \tag{23}$$

Equation (23) says that the power spectrum of the output of the system in Figure 9.2 is just $\psi(\omega)$, as modified by the low-pass filter. In turn, $\psi(\omega)$ is $\phi_{MM}(\omega)$ convolved with itself, where $\phi_{MM}(\omega)$ is the low-pass version of the band-pass power spectrum of the input. In order to derive this result, it was necessary to assume explicitly that $\{X(\cdot)\}$ is a Gaussian process.

In the analysis of the conventional A.M. system using deterministic signals, it turned out that the output signal contained a term that was a replica of the input signal, plus a quadratic distortion term. [Recall (3)]. However, in (23), there is no term that is an exact replica of the input spectrum. The output consists entirely of "quadratic distortion." The reason for this is that there was nothing corresponding to the unity bias term in (1) present in (5).

There is no way to introduce such a term into (5) without causing the process $\{X(\cdot)\}$ to have a time-varying mean, which would make it nonstationary. As we have already pointed out, if we relinquish the assumption of stationarity, then we have to relinquish the concept of spectrum also. Since it was specifically the spectrum of the output that was our original goal, to make an assumption that nullifies the concept of spectrum defeats our whole purpose.

The only way to do a practical analysis and still remain consistent with theoretical principles would be to assume that the input to the square-law device is the sum of *both* a deterministic component of the form (1) plus a stationary, band-pass stochastic process. After carrying through the analysis, we would find that the output could be broken into three parts:

(a) A purely deterministic component.

(b) A purely stationary stochastic component.

(c) A component that is the product of two factors: one factor of the form (a), one factor of the form (b).

If the deterministic part of the input is assumed to be periodic, then conventional Fourier series can determine the spectrum of (a). An analysis of the sort just completed can determine the spectrum of (b). For (c), it is necessary to resort to heuristic reasoning and common sense in order to find an interpretation of the meaning of "spectrum." Further discussion of this point is inappropriate, since the whole problem was only intended to serve as an example of stationary processes in a nonlinear system.

Let us examine the probability distribution of the $\{Y(\cdot)\}$ process next. Define the density function $f_X(x)$ by

$$f_X(x)\,dx = P\{x \le X(t) < x + dx\} \tag{24}$$

If $\{X(\cdot)\}$ is Gaussian and stationary, with zero mean and autocovariance $c_{XX}(\tau)$, then

$$f_X(x) = \frac{1}{\sqrt{2\pi c_{XX}(0)}} e^{-x^2/2c_{XX}(0)}, \qquad -\infty < x < +\infty \tag{25}$$

Also define $f_Y(y)$ by

$$f_Y(y)\,dy = P\{y \le X^2(t) < y + dy\} \tag{26}$$

From elementary probability theory, the law of conservation of probability states, for any real u

$$\int_0^{u^2} f_Y(y)\,dy = \int_{-u}^{u} f_X(x)\,dx \tag{27}$$

which implies

$$f_Y(y) = \frac{1}{\sqrt{2\pi c_{XX}(0)\,y}} e^{-y/2c_{XX}(0)}, \qquad 0 \le y < +\infty \tag{28}$$

This is called a *chi-squared* distribution.

Let $h(t)$ be the impulse response of the low-pass filter. Then, according to Figure 9.2,

$$Z(t) = \int_0^{\infty} h(u)Y(t-u)\,du \tag{29}$$

Since $\{Y(\cdot)\}$ is not a Gaussian process, its probability distribution law will *not* be preserved on passage through a linear system. Therefore, the distribution law for the random variable $Z(t)$ is unknown. The best that can be done is to calculate its moments.

Its mean is

$$E\{Z(t)\} = \int_0^\infty h(u)E\{Y(t-u)\}\,du$$

$$= c_{XX}(0)\int_0^\infty h(u)\,du \tag{30}$$

where we have used (10). The second moment of $Z(t)$ is found as follows:

$$E\{Z^2(t)\} = E\left\{\left(\int_0^\infty h(u)Y(t-u)\,du\right)^2\right\}$$

$$= E\left\{\int_0^\infty\int_0^\infty h(u)h(v)Y(t-u)Y(t-v)\,du\,dv\right\}$$

$$= \int_0^\infty\int_0^\infty h(u)h(v)E\{Y(t-u)Y(t-v)\}\,du\,dv$$

$$= \int_0^\infty\int_0^\infty h(u)h(v)[c_{YY}(v-u) + c_{XX}^2(0)]\,du\,dv \tag{31}$$

Since $\{Y(\cdot)\}$ is a stationary process and the low-pass filter is time invariant, the process $\{Z(\cdot)\}$ is also stationary, so all of its moments are constant. By proceeding as above in a diligent manner, an expression for any desired moment can be obtained.

The Wiener–Volterra Series

Let us introduce this topic by extending the discussion of the square-law detector considered in the previous section. The overall input–output relation for the system depicted in Figure 9.2 may be written

$$Z(t) = \int_{-\infty}^{t} h(t-\tau)X^2(\tau)\,d\tau \tag{32}$$

Although it may seem supererogatory to do so at this point, it is a fact that this relationship may also be written

$$Z(t) = \int_{-\infty}^{t}\int_{-\infty}^{+\infty} h(t-t_1)\delta(t_1-t_2)X(t_1)X(t_2)\,dt_1\,dt_2 \tag{33}$$

Assuming that we are dealing with the impulse response of a causal filter, it will be true that $h(t - t_1) = 0$ for $t_1 > t$. Take this into account and define

$$K_2(t; t_1, t_2) = h(t - t_1)\delta(t_1 - t_2) \tag{34}$$

whence (33) can be rewritten

$$Z(t) = \int_{-\infty}^{\infty}\int_{-\infty}^{\infty} K_2(t; t_1, t_2) X(t_1) X(t_2)\, dt_1\, dt_2 \tag{35}$$

Equation (35) is an example of the input–output relation for what is called a *homogeneous system of degree 2* (see Rugh, 1981). We merely used the square-law detector to illustrate how such an input–output relation can arise in practice.

The generalization of (35) is a homogeneous system of degree n:

$$Z(t) = \int_{-\infty}^{\infty} \cdots \int_{-\infty}^{\infty} K_n(t; t_1, t_2, \ldots, t_n) X(t_1), \ldots, X(t_n)\, dt_1, \ldots, dt_n \tag{36}$$

where we are continuing to denote the input by X and the output by Z.

When a finite collection of such terms is added, the resulting sum is called a polynomial system of degree N:

$$Z(t) = \sum_{n=1}^{N} \int_{-\infty}^{\infty} \cdots \int_{-\infty}^{\infty} K_n(t; t_1, \ldots, t_n) X(t_1), \ldots, X(t_n)\, dt_1, \ldots, dt_n \tag{37}$$

Under appropriate circumstances, the representation can be extended to the case $N \to \infty$. The resulting series is known as a Volterra series or Volterra functional expansion. It is the analog, for dynamical systems, of the Taylor series for functions of a real variable.

We have been using capital letters for the input and output signals, suggesting that they are stochastic processes. Actually, of course, the input–output relationship could well be determined using deterministic signals, and the stochastic problem might be to analyze what happens when stochastic processes are passed through a system having a characterization that has been determined in that way.

Norbert Wiener was a pioneer in the use of the Volterra functional expansion for investigating the passage of stochastic processes through nonlinear systems. When the series is used in this way, it is often referred to as the *Wiener–Volterra series*. More discussion of such applications can be found in Deutsch (1962) and Barrett (1963, 1964). In the remainder of this section, we confine our attention to an introductory example using only deterministic inputs.

In the beginning of this chapter we referred to the fact that nonlinearities are deliberately introduced by the designer into communication systems for certain purposes. Beyond that, however, the system designer also has to contend with the problem of analyzing the performance of systems containing nonlinearities that occur unintentionally or as unmitigatable side effects of design compromises forced upon him or her by considerations of cost.

An example of this latter type of situation is the digital communication link between two ground stations using a satellite repeater. The amplifier used on board the satellite is driven near the saturation point, in order to obtain the maximum power output for a given weight and cost. Another example is found in high-rate digital transmission over telephone channels. The major contribution to the probability of error for such systems operating at rates higher than 4800 bits per second appears to be signal distortion caused by channel nonlinearity. In this case the nonlinearity arises from imperfections in signal compression–expansion equipment commonly used in telephone channels.

Thus, nonlinearities in the transmitter and receiver, specifically modulation and demodulation, are deliberate, while channel nonlinearities are usually unwanted.

In the case of unwanted nonlinearities, their precise analytical representation may be unknown. In that case, the Volterra series is one way to try to model such nonlinearities, and a first task would be to conduct experiments that would permit determination of the kernel functions $K_n(t; t_1, \ldots, t_n)$ in (37).

As an example to illustrate this latter technique, suppose we are confronted with a channel containing a saturation nonlinearity. An equivalent block diagram for *baseband signals* (i.e., after shifting from being centered around some carrier frequency $\omega = \omega_c$ to being centered at $\omega = 0$) is shown in Figure 9.4. The general shape of the saturation nonlinearity is suggestive of the hyperbolic tangent function (tanh), which is plotted in Figure 9.5. A procedure for modeling this channel by using Volterra series would be first to determine the exact Volterra series for the channel with the tanh

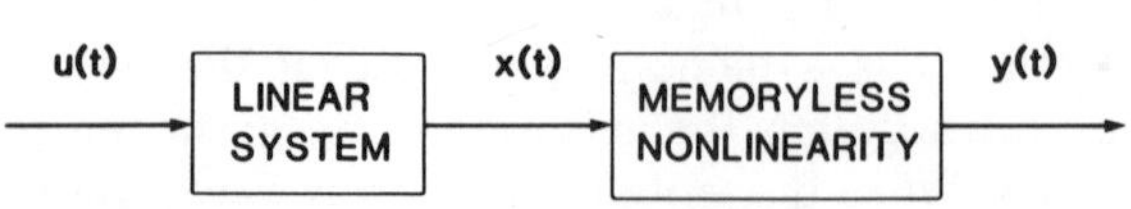

FIGURE 9.4. Model of channel for application of Volterra series.

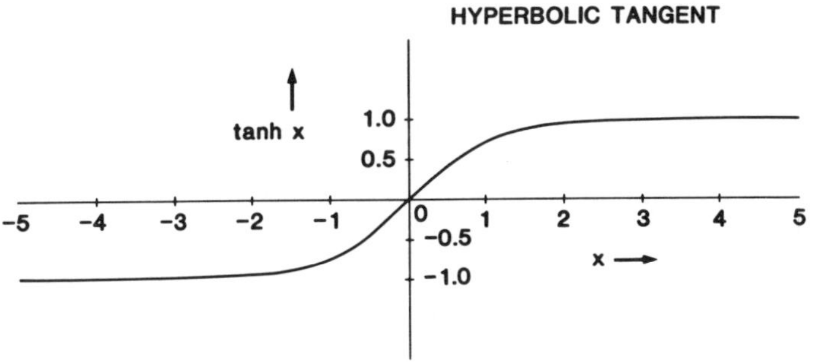

FIGURE 9.5. Plot of hyperbolic tangent, used as model of saturating nonlinearity.

nonlinearity, as a starting point for representing the actual channel. Having done that, one could compare a computer simulation of the tanh channel with experimental data from the actual channel and attempt to modify the expansion coefficients in the tanh channel to bring the simulation into agreement with the experimental data.

Here we content ourselves with deriving the Volterra series for the tanh channel. The expansion for tanh x, carried out to four terms, is

$$\tanh x = x - \frac{x^3}{3} + \frac{2x^5}{15} - \frac{17x^7}{315} + \cdots \tag{38}$$

Referring again to Figure 9.4, the relation between $x(t)$ and $u(t)$ is

$$x(t) = \int_{-\infty}^{\infty} h(t - t_1)u(t_1)\, dt_1 \tag{39}$$

Now we may write

$$\begin{aligned} x^3(t) &= x(t) \cdot x(t) \cdot x(t) \\ &= \left[\int_{-\infty}^{\infty} h(t - t_1)u(t_1)\, dt_1\right]\left[\int_{-\infty}^{\infty} h(t - t_2)u(t_2)\, dt_2\right] \\ &\quad \times \left[\int_{-\infty}^{\infty} h(t - t_3)u(t_3)\, dt_3\right] \\ &= \int_{-\infty}^{\infty}\int_{-\infty}^{\infty}\int_{-\infty}^{\infty} h(t - t_1)h(t - t_2)h(t - t_3) \\ &\quad \times u(t_1)u(t_2)u(t_3)\, dt_1\, dt_2\, dt_3 \end{aligned} \tag{40}$$

Proceeding in this way we arrive at

$$\tanh[x(t)] = \int_{-\infty}^{\infty} h(t-t_1)u(t_1)\,dt_1 - \frac{1}{3}\int_{-\infty}^{\infty}\int_{-\infty}^{\infty}\int_{-\infty}^{\infty} h(t-t_1)h(t-t_2)$$
$$\times h(t-t_3)u(t_1)u(t_2)u(t_3)\,dt_1\,dt_2\,dt_3$$
$$+\frac{2}{15}\underbrace{\int_{-\infty}^{\infty}\cdots\int_{-\infty}^{\infty}}_{5}\prod_{j=1}^{5} h(t-t_j)u(t_j)\,dt_j$$
$$-\frac{17}{315}\underbrace{\int_{-\infty}^{\infty}\cdots\int_{-\infty}^{\infty}}_{7}\prod_{j=1}^{7} h(t-t_j)u(t_j)\,dt_j + \cdots \tag{41}$$

For this example, the kernal functions $K_n(t; t_1, \dots, t_n)$ for even n are all zero. For odd n, the first four are

$$K_1(t; t_1) = h(t-t_1)$$
$$K_3(t; t_1, t_3) = -\tfrac{1}{3}h(t-t_1)h(t-t_2)h(t-t_3)$$
$$K_5(t; t_1, \dots, t_5) = \frac{2}{15}\prod_{j=1}^{5} h(t-t_j) \tag{42}$$
$$K_7(t; t_1, \dots, t_7) = -\frac{17}{315}\prod_{j=1}^{7} h(t-t_j)$$

The analytic representation of this channel, at least formally, by Volterra series, is thus

$$y(t) = \sum_{n=1}^{\infty}\int_{-\infty}^{\infty}\cdots\int_{-\infty}^{\infty} K_n(t; t_1, \dots, t_n)\prod_{j=1}^{n} u(t_j)\,dt_j \tag{43}$$

We must now consider the limitations that need to be placed on the use of this representation to ensure its validity. The question of convergence of Taylor series is, of course, part of the subject of analytic functions of a complex variable. If z is a complex variable, then the function $\tanh z$ is analytic in a disc in the complex z-plane centered at $z = 0$ having a radius of $\pi/2$. Thus, the Taylor series for $\tanh z$ has a domain of convergence of

$|z| < \pi/z$. Hence, even for real-valued arguments x, we must impose the restriction $|x| < \pi/2$ to achieve convergence.

In our present application, this means we must restrict the admissible inputs to some suitable class that will guarantee that this restriction is not violated. Let us investigate to see whether the admissible class of inputs is reasonable for applications. In order to do that, we need to know explicitly the impulse response of the linear part of the channel. Let us suppose that $h(t) = e^{-t}U(t)$, where $U(t)$ is the unit step function.

The relation (39) may be written in the alternate form

$$x(t) = \int_{-\infty}^{\infty} h(t_1)u(t - t_1)\, dt_1 \tag{44}$$

Substituting for $h(t_1)$,

$$x(t) = \int_{0}^{\infty} e^{-t_1}u(t - t_1)\, dt_1 \tag{45}$$

By the triangle inequality for integrals

$$|x(t)| < \int_{0}^{\infty} e^{-t_1}|u(t - t_1)|\, dt_1 \tag{46}$$

since $e^{-t_1} \geq 0$ for all t_1. Suppose we define the admissible class of inputs to be the set of all functions on $(-\infty, \infty)$ bounded in magnitude by a given real number A:

$$|u(t)| \leq A, \qquad -\infty < t < +\infty \tag{47}$$

Substituting this into (46) yields

$$|x(t)| < A\int_{0}^{\infty} e^{-t_1}\, dt_1 \tag{48}$$

and since the value of the integral is itself unity, (48) reduces to

$$|x(t)| < A \tag{49}$$

We have shown that if the linear portion of the channel is given by (45), then (47) implies (49). Hence in that case, the Taylor series will converge provided we set $A < \pi/2$. This restriction might well be reasonable for a class of phase-modulated signals, such as $s(t)$ in equation (59) of Chapter 8.

Summarizing our conclusions now, the Volterra series representation (41) could be used for an analytical investigation of the channel depicted in

Figures 9.4 and 9.5, provided we restrict the inputs to the class defined by (47). This class would include phase-modulated inputs, for example. One's first reaction to this result might be scornful, since we are restricting the class of inputs to a saturation nonlinearity to those signals that are already bounded. However, our conclusion is actually nontrivial, because of the linear portion of the channel between the inputs $u(t)$ and the nonlinearity.

We might proceed, as indicated previously, by truncating the series (41) after a finite number of terms and simulating the result on the computer. The computer simulation could then guide us in modifying the values of the numerical coefficients in front of the integrals, to achieve a better fit to experimental data.

Modulation, Complex Waveforms, and Analytic Signals

We again study a topic that is concerned with the properties of deterministic signals rather than stochastic processes. The justification for including this topic in this book is, as indicated previously, that a first-year graduate course on stochastic processes, for communications systems majors in electrical engineering, traditionally includes some discussion of signal theory. The topic about to be addressed here is part of this classical material. It concerns a formalism for dealing with band-pass signals.

Signals that are purely real-valued possess Fourier transforms having certain symmetry properties around zero frequency. In particular, the real part of the transform must be an even function of frequency, whereas the imaginary part of the transform must be a purely odd function. Consequently, the entire signal can be reconstructed from a knowledge of its spectrum for positive frequencies only.

We begin by developing some properties of real-valued signals and their transforms. Let $y(t)$ be a real signal and let $Y(i\omega)$ be its Fourier transform.

Consider a physically impossible, but mathematically legitimate, hypothetical linear system having a transfer function

$$A(i\omega) = \begin{cases} 2, & \omega > 0 \\ 0, & \omega < 0 \end{cases} \tag{50}$$

In terms of the unit step function, $A(i\omega) = 2U(\omega)$. Let $a(t)$ be the corresponding impulse response, that is, the inverse Fourier transform of $A(i\omega)$. We deal with Fourier transforms here rather than Laplace, because we want to allow time functions to be defined for $-\infty < t < +\infty$. Explicitly,

$$a(t) = \frac{1}{\pi} \int_0^\infty e^{i\omega t} \, d\omega \tag{51}$$

Arguments based on the theory of distributions lead to the conclusion that

$$a(t) = \delta(t) + \frac{i}{\pi t} \tag{52}$$

If the signal $y(t)$ with spectrum $Y(i\omega)$ is passed through this hypothetical linear system, the output spectrum is $2Y(i\omega)U(\omega)$. By the convolution theorem, from (52), the output time function must be $y(t) + i\hat{y}(t)$, where

$$\hat{y}(t) = \frac{1}{\pi}\int_{-\infty}^{\infty} \frac{y(\theta)}{t - \theta}\, d\theta \tag{53}$$

is called the *Hilbert transform* of $y(t)$. Elsewhere in this book we use the circumflex to denote the optimal estimate of a time signal. In this section only, that notation means Hilbert transform, not optimal estimate, because this particular notation is standardly used by the experts in this area.

Because of the singularity at $\theta = t$ in the integrand of (53), the integral has to be interpreted as the Cauchy principal value.

We will adopt the notation of using a lowercase Greek psi, ψ, with the appropriate subscript, for denoting the output time function from the linear system with transfer function $A(i\omega)$. In this case,

$$\psi_Y(t) = y(t) + i\hat{y}(t) \tag{54}$$

The function $\psi_Y(t)$ is called the *analytic signal* or *pre-envelope* corresponding to $y(t)$. Its Fourier transform is denoted by $\Psi_Y(i\omega)$.

Now let $y(t)$ be a narrow-band band-pass signal and let $z(t)$ be a narrow-band baseband signal. Specifically, if $Y(i\omega)$ is the Fourier transform of $y(t)$, then we are assuming that the effective support of $Y(i\omega)$ is the set

$$[-\omega_c - \Omega, -\omega_c + \Omega] \cup [\omega_c - \Omega, \omega_c + \Omega]$$

Further, if $Z(i\omega)$ is the Fourier transform of $z(t)$, then the effective support of $Z(i\omega)$ is the set $[-\omega_L, \omega_L]$. Therefore, if $\psi_Y(t)$ and $\psi_Z(t)$ are the respective analytic signals corresponding to $y(t)$ and $z(t)$, with respective Fourier transforms $\Psi_Y(i\omega)$ and $\Psi_Z(i\omega)$, then the effective support of $\Psi_Y(i\omega)$ is $[\omega_c - \Omega, \omega_c + \Omega]$ and the effective support of $\Psi_Z(i\omega)$ is $[0, \omega_L]$. We are specifically assuming that

$$2\omega_L \ll \omega_c - \Omega \qquad \Omega \ll \omega_c \tag{55}$$

We want to investigate the spectrum $X(i\omega)$ of the product

$$x(t) = z(t)y(t) \tag{56}$$

Consider the product $z(t)\psi_Y(t)$. By definition

$$\Psi_Y(i\omega) = 2Y(i\omega)U(\omega) \tag{57}$$

The Fourier transform of $z(t)\psi_Y(t)$ is the convolution in the frequency domain of $Z(i\omega)$ with $\Psi_Y(i\omega)$. Because of the assumptions (55), the spectrum $Z(i\omega) * \Psi_Y(i\omega)$ still vanishes for all $\omega < 0$.

In order to recover a signal from its analytic representation, it is clear from (54) that we need only take the real part. Thus

$$y(t) = \text{Re}[\psi_Y(t)] \tag{58}$$

Therefore we may write (56) as

$$x(t) = z(t)\,\text{Re}[\psi_Y(t)] \tag{59}$$

Since $z(t)$ is purely real, then of course $x(t)$ is purely real. If $\hat{x}(t)$ is its Hilbert transform, then the analytic signal $\psi_X(t)$ is given by

$$\psi_X(t) = x(t) + i\hat{x}(t) \tag{60}$$

Let $\mathscr{H}[\cdot]$ denote "Hilbert transform of." Now

$$\hat{y}(t) = \mathscr{H}[y(t)] \tag{61}$$

$$\hat{x}(t) = \mathscr{H}[x(t)] \tag{62}$$

In general, it is *false* that

$$\mathscr{H}[z(t)y(t)] = z(t)\mathscr{H}[y(t)] \tag{63}$$

However, *if* the spectra of the various signals obey the restrictions (55), *then* (63) is a true equation. In *that* case,

$$\begin{aligned} z(t)\psi_Y(t) &= z(t)y(t) + iz(t)\hat{y}(t) \\ &= z(t)y(t) + i\widehat{[z(t)y(t)]} \\ &= x(t) + i\hat{x}(t) \\ &= \psi_X(t) \end{aligned} \tag{64}$$

Approaching the situation another way, we noted in the discussion connected with (57) that the convolution $Z(i\omega) * \Psi_Y(i\omega)$ vanishes for $\omega < 0$. In that case, its inverse Fourier transform

$$\mathscr{F}^{-1}[Z(i\omega) * \Psi_Y(i\omega)]$$

must be an analytic signal. The question is, what is its real part?

The statement in the frequency domain that corresponds to (63) concerns the properties of the Fourier transform of the product $z(t)y(t)$. Now, this transform is the convolution

$$Z(i\omega) * Y(i\omega)$$

Specifically

$$X(i\omega) = \mathscr{F}[x(t)] = Z(i\omega) * Y(i\omega) \tag{65}$$

and by definition

$$\Psi_X(i\omega) = 2X(i\omega)U(\omega) \tag{66}$$

The statement in the frequency domain that corresponds to (63) is

$$U(\omega)[Z(i\omega) * Y(i\omega)] = Z(i\omega) * [Y(i\omega)U(\omega)] \tag{67}$$

Once again: (67) holds *only* if (55) is met.

Comparing (65), (66), and (67) and recalling (57), we have, when (55) holds,

$$\Psi_X(i\omega) = Z(i\omega) * \Psi_Y(i\omega) \tag{68}$$

Evidently, (68) is the statement in the frequency domain which corresponds to the time domain statement (64).

A special case of the above result is the case $y(t) = \cos \omega_c t$, so that

$$x(t) = z(t)\cos \omega_c t \tag{69}$$

In this case $\Omega = 0$, so (55) reduces to

$$2\omega_L < \omega_c \tag{70}$$

In that case

$$Y(i\omega) = \pi[\delta(\omega - \omega_c) + \delta(\omega + \omega_c)] \tag{71}$$

$$\Psi_Y(i\omega) = 2\pi\delta(\omega - \omega_c) \tag{72}$$

$$Z(i\omega) * \Psi_Y(i\omega) = \frac{1}{2\pi}\int_{-\infty}^{\infty} \Psi_Y(i\omega')Z(i\omega - i\omega')\, d\omega'$$

$$= \int_{-\infty}^{\infty} Z(i\omega - i\omega')\delta(\omega' - \omega_c)\, d\omega' = Z(i\omega - i\omega_c) \tag{73}$$

Therefore in this case, the spectrum of the analytic signal $\psi_X(t)$ corresponding to $x(t) = z(t)\cos\omega_c t$ is

$$\Psi_X(i\omega) = Z(i\omega - i\omega_c) \tag{74}$$

that is, just the baseband spectrum shifted over so that it is centered about the carrier frequency ω_c. In the problems section, we will investigate the interpretation of the A.M. transmitter given at the beginning of this chapter in terms of analytic signals and their spectra, using this result.

However, the Hilbert transform of $\cos\omega_c t$ is just $\sin\omega_c t$, so that

$$\psi_Y(t) = \cos\omega_c t + i\sin\omega_c t = e^{i\omega_c t} \tag{75}$$

and of course

$$y(t) = \text{Re}[\psi_Y(t)] = \cos\omega_c t \tag{76}$$

as asserted by (58). Then, as given by (64),

$$\begin{aligned}\psi_X(t) &= z(t)\psi_Y(t) = z(t)e^{i\omega_c t}\\ &= x(t) + i\hat{x}(t) = z(t)\cos\omega_c t + iz(t)\sin\omega_c t\end{aligned} \tag{77}$$

Equation (63) in this case reduces to

$$\mathscr{H}[z(t)\cos\omega_c t] = z(t)\sin\omega_c t \tag{78}$$

which is valid provided (70) holds. The relation $x(t) = \text{Re}[\psi_X(t)]$ in this case is just

$$x(t) = z(t)\cos\omega_c t = \text{Re}[z(t)e^{i\omega_c t}] \tag{79}$$

The Radar Uncertainty Principle

Having thus surveyed some of their properties, we now illustrate the application of analytic signals by introducing a concept that is important in radar design: the *radar ambiguity function*. A radar system basically functions by transmitting a periodic sequence of narrow band signals and scanning the environment for return echoes. By comparing the waveform of a received echo with the known waveform that was transmitted, it is possible not only to detect the presence of targets in the vicinity of the radar installation, in the direction of the beam formed by the antenna, but also to determine their range (distance) from the antenna and the velocity with which they are moving relative to the antenna.

A problem of concern to radar designers is to select the most appropriate waveform for transmission. It is important to be able to discriminate between two different targets that are in only slightly different locations. It is also important to be able to resolve the difference in velocity between two such targets that are moving with slightly different velocities. Not all waveforms have equally good resolution properties. Some waveforms are better at resolving range; others blur range but discriminate sharply in velocity. It turns out that good range resolution requires concentration in time, while good velocity resolution requires concentration in frequency.

There is a fundamental limitation on the performance of any radar system in regard to its ability to achieve simultaneously high resolution in *both* position and velocity. Because of a certain analogy with a similar phenomenon that occurs in quantum mechanics, this limitation has come to be known as the *Radar Uncertainty Principle*. It is closely related to the Time–Bandwidth Uncertainty Principle we discussed at the end of Chapter 8. However, because of the specialized form that it takes in the radar application, it is of independent interest.

The ideal radar pulse would be of short duration in time and simultaneously concentrated in a narrow band of frequencies. Actually, this is not an accurate statement. Because of peak power limitations on the radar transmitter, such a pulse in fact creates other difficulties in system design. In Chapter 6 we showed that it is possible to define a time autocorrelation function for deterministic, finite energy signals and to use its Fourier transform in a way analogous to the use of the power spectral density of a stationary random process, to gauge the concentration of energy in various portions of the frequency spectrum. Having recalled that, we may say that a more accurate statement concerning the ideal radar pulse is that the pulse itself may be somewhat spread in time, in order to meet peak power limitations at the transmitter, but its *autocorrelation function* should be effectively supported on a very short time interval, while simultaneously possessing a Fourier transform that is highly concentrated in frequency.

Some of the pulses used in radar systems might be described as sinusoidal waveforms that have been subjected to modulation in *both* amplitude and frequency. If $x(t)$ is such a signal, its factorization as in (56) may not be unique: there may be more than one choice of $y(t)$ and $z(t)$. It is popular to use a representation for $x(t)$ which resembles (79). However, because of the drift in instantaneous frequency which is deliberately introduced, the choice of ω_c is somewhat arbitrary. It has been found useful to choose ω_c somewhat as the "center of gravity" of the spectrum of $\Psi_x(i\omega)$. In order to use a representation of the form (79), it then becomes expedient to allow the envelope $z(t)$ to be complex:

$$z(t) = u(t) + iv(t) \tag{80}$$

As before, $z(t)$ is *not* an analytic signal, because here $v(t)$ is *not* the Hilbert transform of $u(t)$. Moreover, the spectrum of $Z(i\omega)$ still contains negative frequencies as well as positive ones. Of course now the relationship (69) no longer holds, although (77) does continue to hold, and $\psi_X(t)$ *is* still an analytic signal, provided $z(t)$ remains low pass.

What does still hold in particular is the statement

$$x(t) = \mathrm{Re}[z(t)e^{i\omega_c t}] \tag{81}$$

which in the frequency domain still implies the validity of the equation (74). In fact, since $\hat{x}(t)$ can be found directly from $x(t)$ by the Hilbert transform (53), or, equivalently, since $\Psi_x(i\omega)$ can be found directly from $X(i\omega)$ by truncating it at $\omega = 0$, as in (57), then (74) can be taken as the *definition* of $Z(i\omega)$ once ω_c is selected.

When this approach is adopted, the (now complex-valued) baseband signal $z(t)$ is called the *complex envelope* of the signal $x(t)$. Because of the known desirable features of a radar pulse, we may still assume that $Z(i\omega)$ is effectively supported on some baseband interval $[-\omega_L, \omega_L]$ where ω_L satisfies (70).

Since $Z(i\omega)$ is the Fourier transform of $z(t)$,

$$Z(i\omega) = \int_{-\infty}^{\infty} z(t)e^{-i\omega t}\,dt \tag{82}$$

then by the Parseval identity

$$\mathscr{E} = \frac{1}{2\pi}\int_{-\infty}^{\infty} |Z(i\omega)|^2\,d\omega = \int_{-\infty}^{\infty} |z(t)|^2\,dt \tag{83}$$

For our purpose of presenting the Radar Uncertainty Principle, it is helpful to consider $|z(t)|^2$ and $(1/2\pi)|Z(i\omega)|^2$ as analogous to probability density functions. In order to pursue this, let us assume both sides of (83) are equal to unity, that is, assume the signals are scaled so that this happens. In that case, let us adopt the second moments of these two "distributions" as measures of the respective spread or dispersion in time and frequency:

$$D_T = \int_{-\infty}^{\infty} t^2|z(t)|^2\,dt \tag{84}$$

$$D_F = \frac{1}{2\pi}\int_{-\infty}^{\infty} \omega^2|Z(i\omega)|^2\,d\omega \tag{85}$$

Now consider the following object:

$$R(\tau, \beta) = \int_{-\infty}^{\infty} z(t) z^*(t + \tau) e^{-i\beta t} \, dt \tag{86}$$

Equation (86) is the definition of the symbol $R(\tau, \beta)$. Note these properties:

$$\frac{1}{2\pi} \int_{-\infty}^{\infty} R(0, \beta) e^{i\beta t} \, d\beta = |z(t)|^2 \tag{87}$$

$$\frac{1}{2\pi} \int_{-\infty}^{\infty} R(\tau, 0) e^{i\omega\tau} \, d\tau = \frac{1}{2\pi} |Z(i\omega)|^2 \tag{88}$$

Therefore, $R(\tau, \beta)$ may be considered a kind of joint characteristic function. In keeping with that interpretation, we see that the two dispersions D_T and D_F defined in (84) and (85) may be written

$$D_T = \left[\frac{-\partial^2}{\partial \beta^2} R(0, \beta) \right]_{\beta=0} \tag{89}$$

$$D_F = \left[-\frac{\partial^2}{\partial \tau^2} R(\tau, 0) \right]_{\tau=0} \tag{90}$$

Equation (90) is easier to verify if one first shows that from (86) it follows that $R(\tau, \beta)$ may also be represented as

$$R(\tau, \beta) = \frac{1}{2\pi} \int_{-\infty}^{\infty} Z^*(i\omega) Z(i\omega + i\beta) e^{-i\omega\tau} \, d\omega \tag{91}$$

Theorem (The Radar Uncertainty Principle). Consider the class of all complex-valued waveforms $z(\cdot)$ normalized so that $\mathscr{E} = 1$ as defined by (83) and for which D_T and D_F as defined, respectively, by (84) and (85) are both finite. No matter how the waveform is chosen, so long as it is in this class, the product $D_T D_F$ is bounded below:

$$\boxed{D_T D_F \geq \tfrac{1}{4}} \tag{92}$$

PROOF. Applying (90) to (86), one finds

$$D_F = -\int_{-\infty}^{\infty} z(t) \frac{d^2 z^*(t)}{dt^2} = \int_{-\infty}^{\infty} z'(t) z'^*(t) \, dt \tag{93}$$

where $z'(t) = (d/dt)z(t)$, and we have integrated by parts to obtain the second equation.

Now consider the following identity:

$$\frac{d}{dt} zz^* = z'z^* + zz'^* \tag{94}$$

We have suppressed the argument for brevity. Multiply both sides by t and integrate over the whole line:

$$\int_{-\infty}^{\infty} t \frac{d}{dt} zz^* \, dt = \int_{-\infty}^{\infty} tz'z^* \, dt + \int_{-\infty}^{\infty} tzz'^* \, dt \tag{95}$$

Apply the triangle inequality:

$$\left| \int_{-\infty}^{\infty} t \frac{d}{dt} zz^* \, dt \right| \le \left| \int_{-\infty}^{\infty} tz'z^* \, dt \right| + \left| \int_{-\infty}^{\infty} tzz'^* \, dt \right| \tag{96}$$

The integral on the left may be evaluated by integration by parts:

$$\int_{-\infty}^{\infty} t \frac{d}{dt} zz^* \, dt = tzz^* \Big|_{-\infty}^{\infty} - \int_{-\infty}^{\infty} zz^* \, dt \tag{97}$$

By the assumption that D_T in (84) is finite, the first term must vanish. By the assumption that $\mathscr{E} = 1$ in (83), we conclude that the absolute value of the integral on the left side of (96) is equal to unity.

The two integrals on the right of (96) are bounded by the Schwarz inequality:

$$\left| \int_{-\infty}^{\infty} tz'z^* \, dt \right| \le \left[\left(\int_{-\infty}^{\infty} t^2 |z^*|^2 \, dt \right) \left(\int_{-\infty}^{\infty} |z'|^2 \, dt \right) \right]^{1/2} \tag{98}$$

$$\left| \int_{-\infty}^{\infty} tzz'^* \, dt \right| \le \left[\left(\int_{-\infty}^{\infty} t^2 |z|^2 \, dt \right) \left(\int_{-\infty}^{\infty} |z'^*|^2 \, dt \right) \right]^{1/2} \tag{99}$$

The quantities on the right side of both (98) and (99) are equal to each other, and their common value, from (84) and (93), is $\sqrt{D_T D_F}$. Using all of these results in (96) yields the inequality

$$1 \le 2\sqrt{D_T D_F} \tag{97}$$

which upon rearrangement is (92). This concludes the proof.

The interpretation of this theorem is similar to the uncertainty principle given in Chapter 8: it is not possible to find a waveform that concentrates energy simultaneously in time and frequency beyond certain fixed bounds. The consequence of this for radar design is that there is an inherent fundamental limit on the ability of any radar system to resolve two targets simultaneously in position and velocity. An increase in position resolution can be bought only at the price of a sacrifice of velocity resolution and vice versa.

The squared magnitude of the object $R(\tau, \beta)$ defined in (86), or equivalently (91), is sometimes called the *Radar Ambiguity Function*. A study of its behavior for a particular system is often helpful to the system designer for gaining insights into system performance. Its properties are discussed in detail in Cook and Bernfeld (1967), where further references may be found. More information concerning radar signal design may also be found in Burdic (1968), Franks (1981), and Vakman (1968).

Problems

1. Verify that when (15) is used in (16), the result is (17).

2. Use (30) and (31) to show that the variance of the random variable $Z(t)$ is

$$\sigma_Z^2 = \int_0^\infty \int_0^\infty c_{YY}(u - v)h(u)h(v)\,du\,dv$$

3. (Rugh) Consider a Volterra system of the form

$$y(t) = \sum_{n=1}^{\infty} \int_0^t h_n(t - \sigma_1, \ldots, t - \sigma_n)u(\sigma_1) \cdots u(\sigma_n)\,d\sigma_1 \cdots d\sigma_n$$

where $h_1(t) = U(t)$ (unit step), and for $n > 1$,

$$h_n(t_1, \ldots, t_n) = a_n\delta(t_1 - t_2)\delta(t_2 - t_3) \cdots \delta(t_{n-1} - t_n)$$

The integral sign is to be interpreted as an integral of appropriate multiplicity, with all limits from 0 to t. Give conditions under which this series converges. Give an alternative form for the input–output relationship.

4. Give an argument to verify (51) and (52). Specifically, define for real $\alpha \geq 0$

$$a(t, \alpha) = \frac{1}{\pi} \int_0^\infty e^{-\alpha\omega} e^{i\omega t}\,d\omega$$

Evaluate the integral explicitly to obtain

$$a(t, \alpha) = \frac{1}{\pi} \frac{1}{\alpha - it}$$

Now write

$$\frac{1}{\alpha - it} = \frac{\alpha + it}{(\alpha - it)(\alpha + it)} = \frac{\alpha + it}{\alpha^2 + t^2}$$

Evidently

$$\lim_{\alpha \to 0} \frac{it}{\alpha^2 + t^2} = \frac{i}{t}$$

Your problem is to show that

$$\lim_{\alpha \to 0} \frac{\alpha}{\alpha^2 + t^2} = \pi\delta(t)$$

If $f(t)$ is any bounded, continuous function defined for $-\infty < t < +\infty$, it is equivalent to prove

$$\lim_{\alpha \to 0} \int_{-\infty}^{\infty} \frac{\alpha}{\alpha^2 + t^2} f(t)\, dt = \pi f(0)$$

5. Verify the statement just before (75) that $\mathscr{H}[\cos \omega_c t] = \sin \omega_c t$. Specifically, show that

$$\frac{1}{\pi} \int_{-\infty}^{\infty} \frac{\cos \omega_c \theta}{t - \theta}\, d\theta = \sin \omega_c t$$

Note that for $t = 0$ this says that

$$\frac{1}{\pi} \int_{-\infty}^{\infty} \frac{\cos \omega_c \theta}{-\theta}\, d\theta = 0$$

However, it is a fact that

$$\int_0^{\infty} \frac{\cos \omega_c \theta}{\theta}\, d\theta = \infty$$

Let $\alpha > 0$. Then $\int_\alpha^\infty [(\cos \omega_c \theta)/\theta]\, d\theta < \infty$. Show that $\int_{-\infty}^{-\alpha} [(\cos \omega_c \theta)/\theta]\, d\theta = -\int_\alpha^\infty [(\cos \omega_c \theta)/\theta]\, d\theta$ so that

$\int_{-\infty}^{-\alpha}[(\cos \omega_c\theta)/\theta]\, d\theta + \int_{\alpha}^{\infty}[(\cos \omega_c\theta)/\theta]\, d\theta = 0$ for all values of $\alpha > 0$. Hence

$$\lim_{\alpha \to 0}\left[\int_{-\infty}^{-\alpha} \frac{\cos \omega_c\theta}{-\theta}\, d\theta + \int_{\alpha}^{\infty} \frac{\cos \omega_c\theta}{-\theta}\, d\theta\right] = 0$$

In the same way you need to show that

$$\lim_{\alpha \to 0}\left[\frac{1}{\pi}\int_{-\infty}^{t-\alpha} \frac{\cos \omega_c\theta}{t-\theta}\, d\theta + \frac{1}{\pi}\int_{t+\alpha}^{\infty} \frac{\cos \omega_c\theta}{t-\theta}\, d\theta\right] = \sin \omega_c t$$

This is known as the *Cauchy principal value of an improper integral.*

6. Refer to the A.M. communication system depicted in Figure 9.1. Consider only the portion labeled "modulator and transmitter," with input $m(t)$ and output $x(t)$. If the Fourier transform of $m(t)$ is $M(i\omega)$, find the spectrum $\Psi_x(i\omega)$ of the analytic signal $x(t) + i\hat{x}(t)$.

7. Let $z(\cdot)$ be a complex-valued member of the complex Hilbert space $L_2(-\infty, -\infty)$, where

$$\|z\|^2 = \int_{-\infty}^{\infty} z(t)z^*(t)\, dt$$

Define $R_Z(\tau, \beta)$ as in (86):

$$R_Z(\tau, \beta) = \int_{-\infty}^{\infty} z(t)z^*(t+\tau)e^{-i\beta t}\, dt$$

Demonstrate the following properties:

(a) $|R_Z(\tau, \beta)|^2 = |R_Z(-\tau, -\beta)|^2$

(b) $|R_Z(\tau, \beta)|^2 \le |R_Z(0,0)|^2$

(c) $|R_Z(0,0)| = \|z\|^2$

(d) $\int_{-\infty}^{\infty}\int_{-\infty}^{\infty}|R_Z(\tau, \beta)|^2\, d\tau\, d\beta = 2\pi\|z\|^4$

Property (d) shows that if $|R_Z(\tau, \beta)|^2$ is interpreted as an unnormalized density of ambiguity in the τ–β plane, the total volume under this surface is always the same for all signals of equal total energy. Therefore, the ambiguity can only be redistributed in various ways, much like an obese person who exercises and changes shape without losing weight.

8. Use (84) through (86) to demonstrate (87) through (91).

9. Let $z(t)$ be a Gaussian waveform:

$$z(t) = Ke^{-\alpha t^2}, \qquad \alpha > 0$$

Let K be chosen so that

$$\mathscr{E} = \int_{-\infty}^{\infty} |z(t)|^2 \, dt = 1$$

Compute $Z(i\omega)$, the Fourier transform of $z(t)$ and show that this waveform achieves the lower bound $D_T D_F = \frac{1}{4}$ in the Radar Uncertainty Principle (92), for arbitrary $\alpha > 0$.

Chapter 10

Nonstationary Random Sequences

Scalar-Valued Sequences

The objects of study in this chapter are singly infinite random sequences, defined only on the non-negative integers. The statistics of the members of these sequences are permitted to vary with time. We begin by examining scalar-valued sequences, and then generalize to the consideration of vector-valued sequences in order to introduce state-space methods of analysis.

The first such sequence we require is the white Gaussian noise (WGN) sequence $\{V(\cdot)\}$. The successive members $V(0), V(1), V(2), \ldots,$ are mutually independent, zero mean Gaussian random variables. We define unit WGN to be the sequence for which all members have variance one. Let us use this sequence as the forcing term on a first-order scalar linear difference equation and study the resulting output. Linear difference equations are standard models for generating stochastic processes having various properties in practice, and this simple one will be a prototype for what we will study in the remainder of this chapter.

Consider now

$$X(t+1) = aX(t) + V(t), \qquad t = 0, 1, 2, \ldots \tag{1}$$

where a is a constant. In words, (1) says that the present value of X is the sum of the present forcing term and a weighted memory of the previous value of X. In order to start the procedure, we must be given the initial value of X, in this case $X(0)$. Since we are dealing with stochastic processes and not deterministic sequences, this initial condition is itself random. We will assume $X(0)$ is a Gaussian r.v., independent of all of the $V(t)$, with mean m and covariance c.

Equation (1) is most directly solved by recursion, as follows:

$$\begin{aligned}
X(1) &= aX(0) + V(0) \\
X(2) &= aX(1) + V(1) = a[aX(0) + V(0)] + V(1) \\
&= a^2X(0) + aV(0) + V(1) \qquad (2) \\
X(3) &= aX(2) + V(2) = a[a^2X(0) + aV(0) + V(1)] + V(2) \\
&= a^3X(0) + a^2V(0) + aV(1) + V(2)
\end{aligned}$$

From this procedure the general formula suggests itself, which may be verified by direct substitution:

$$X(t) = a^tX(0) + \sum_{k=0}^{t-1} a^{t-k-1}V(k) \qquad (3)$$

Since $X(0), V(0), V(1), V(2), \ldots, V(t-1)$ are all mutually independent Gaussian random variables, it is easy to give a complete probabilistic description of the r.v. $X(t)$. Since the mean of a sum of independent r.v.'s is the sum of the means, and the variance of such a sum is the sum of the variances, we see that

$$E[X(t)] = a^tm \qquad (4)$$

$$E\{[X(t) - a^tm]^2\} = a^{2t}c + \sum_{k=1}^{t} a^{2(t-k)} \qquad (5)$$

If desired, the sum may be written in closed form, by the identity

$$1 + a^2 + (a^2)^2 + \cdots + (a^2)^{t-1} = \frac{(a^2)^t - 1}{a^2 - 1} \qquad (6)$$

Since $X(t)$ is Gaussian, knowledge of the mean and variance is equivalent to knowledge of the distribution function itself. We see now that if $V(t)$ is not assumed Gaussian, but merely to be a sample from a second-order random process, then all we have is the mean and the variance of X at this point. There is no way to make any inference about the distribution function itself.

It is instructive at this point to take the time and effort to consider the properties of the joint probability distribution of all of the random variables $X(0), X(1), X(2), \ldots, X(t)$. Here is one place where the practice of using capital letters to denote random variables serves us well, because we can now use the corresponding lowercase letters to designate the arguments of the probability density function. Therefore, consider the joint density function

$$\begin{aligned} p(x(0), x(1), x(2), \ldots, x(t))\, dx(0)\, dx(1) \cdots dx(t) \\ = P\{x(0) \le X(0) < x(0) + dx(0), x(1) \le X(1) < x(1) + dx(1), x(2) \\ \le X(2) < x(2) + dx(2), \ldots, x(t) \le X(t) < x(t) + dx(t)\} \end{aligned} \tag{7}$$

Let us also consider the joint distribution of the input white noise sequence

$$\begin{aligned} f(v(0), v(1), \ldots, v(t-1))\, dv(0)\, dv(1)\, dv(t-1) \\ = P\{v(0) \le V(0) < v(0) + dv(0), \ldots, v(t-1) \\ \le V(t-1) < v(t-1) + dv(t-1)\} \end{aligned} \tag{8}$$

Since all these r.v.'s are mutually independent,

$$f(v(0), v(1), \ldots, v(t-1)) = f_1(v(0)) f_1(v(1)) \cdots f_1(v(t-1)) \tag{9}$$

where, since this is unit WGN, for every k we have

$$f_1(v(k)) = \frac{1}{\sqrt{2\pi}} e^{-v^2(k)/2} \tag{10}$$

Let us put all of the scalar variables $X(1), X(2), \ldots, X(t)$ into a vector of dimension t. Do the same for $V(0), V(1), \ldots, V(t-1)$. Work out equation (3) for all times from 1 to t, inclusive, and express all of these equations simultaneously in matrix form

$$\begin{bmatrix} X(1) \\ X(2) \\ \vdots \\ X(t) \end{bmatrix} = \begin{bmatrix} a \\ a^2 \\ \vdots \\ a^t \end{bmatrix} X(0) + \begin{bmatrix} 1 & 0 & \cdots & 0 \\ a & 1 & \cdots & 0 \\ a^{t-1} & a^{t-2} & & 1 \end{bmatrix} \begin{bmatrix} V(0) \\ V(1) \\ \vdots \\ V(t-1) \end{bmatrix} \tag{11}$$

The matrix in (11) is lower triangular. It has all 1's on the main diagonal. Every element on the diagonal just below the main diagonal is the constant a. Every element on the diagonal below that is a^2, and so on. Every element in the upper triangle, above the main diagonal, is zero. A matrix having the property that every element on each diagonal is the same, is called a *Toeplitz form*.

By assumption $X(0)$ is Gaussian with mean m and covariance c, so the corresponding density function is

$$p_0(x(0)) = \frac{1}{\sqrt{2\pi c}} e^{-[x(0)-m]^2/2c} \tag{12}$$

Now include the identity $X(0) = X(0)$ in the set of equations in (11). Augment both the vectors $[X(1) \cdots X(t)]^T$ and $[V(0) \cdots V(t-1)]^T$ to include $X(0)$ as the first element. Rewrite (11) in the form

$$\begin{bmatrix} X(0) \\ X(1) \\ X(2) \\ \vdots \\ X(t) \end{bmatrix} = \begin{bmatrix} 1 & 0 & \cdots & 0 \\ a & 1 & \cdots & 0 \\ \vdots & \vdots & & \\ a^t & a^{t-1} & & 1 \end{bmatrix} \begin{bmatrix} X(0) \\ V(0) \\ V(1) \\ \vdots \\ V(t-1) \end{bmatrix} \tag{13}$$

Since the transformation in (13) is clearly nonsingular (the Jacobian determinant is equal to one, by inspection), the standard rules of probability theory show how to find the joint density function $p(x(0), x(1), x(2), \ldots, x(t))$ in (7) from the known joint density $p_0(x(0))f(v(0), v(1), \ldots, v(t-1))$.

In a similar manner, we can repeat this whole process with the time argument increased by one, to obtain the joint density function $p(x(0), x(1), \ldots, x(t), x(t+1))$. By the rules of probability theory, taking the ratio gives us the conditional probability density function

$$p(x(t+1)|x(t), x(t-1), \ldots, x(1), x(0)) = \frac{p(x(0), x(1), \ldots, x(t), x(t+1))}{p(x(0), x(1), \ldots, x(t-1), x(t))} \tag{14}$$

Now, rewriting equation (1),

$$X(t+1) - aX(t) = V(t) \tag{15}$$

The probability density for $V(t)$ is given by equation (10). Therefore, the conditional probability density for $X(t+1)$ given $X(t)$ is necessarily

$$p(x(t+1)|x(t)) = \frac{1}{\sqrt{2}\pi} e^{-[x(t+1)-ax(t)]^2/2} \tag{16}$$

The interesting fact is that if one goes through the effort of carrying out the procedure leading to (14) explicitly, one finds that all of the other variables identically drop out of the expression, leaving

$$p(x(t+1)|x(t),\ldots,x(1),x(0)) = p(x(t+1)|x(t)) \tag{17}$$

where $p(x(t+1)|x(t))$ is given by (16).

The purpose of this somewhat belabored discussion is to take advantage of the context and provide an explicit illustration of what we mean when we say that a process is Markov. Equation (17) will be recognized as the defining relationship for a Markov process, according to the definition given in Chapter 1.

The conceptual content of this equation is that the state of the process at time t summarizes everything about the future of the process that can be known by looking at the entire past. It is worth noting that the difference equation (1) and its solution by recursion in (2), thereby generating the sequence $\{X(t)\}$, is an example of what was called the *successive* viewpoint in Chapter 4. By contrast, equations (11) and (13) represent the *simultaneous* point of view. In order to grasp the full meaning of the definition of a Markov process, represented by equations (14) and (17), it is necessary to have access to both viewpoints.

Vector-Valued Sequences

Now consider the generalization of (1):

$$\begin{aligned} X(t+n) + a_1X(t+n-1) + a_2X(t+n-2) + \cdots + a_nX(t) \\ = b_0V(t+m) + b_1V(t+m-1) + \cdots + b_mV(t) \end{aligned} \tag{18}$$

It seems evident that (18) can also be solved by a suitable recursive procedure, but just a few attempts to do that will convince the reader that finding the solution is basically a messy bookkeeping problem. It would be nice if there were some technique that would make the recursion as easy as (2) was. There is such a technique, involving the use of vector/matrix notation. What we do is rewrite (18) as a first-order vector difference equation. Introduce the vector

$$\mathbf{X}(t) = \begin{bmatrix} X_1(t) \\ X_2(t) \\ \vdots \\ X_n(t) \end{bmatrix} \tag{19}$$

Define the matrix

$$\mathbf{A} = \begin{bmatrix} -a_1 & 1 & 0 & 0 \\ -a_2 & 0 & 1 & 0 \\ \vdots & \cdot & \cdot & \cdot \\ \cdot & \cdot & \cdot & \cdot \\ -a_n & 0 & \cdot & 0 \end{bmatrix} \tag{20}$$

and the vector

$$\mathbf{b} = \begin{bmatrix} 0 \\ \vdots \\ b_0 \\ \vdots \\ b_{m-1} \\ b_m \end{bmatrix} \tag{21}$$

In (18) and (21) we assume $m < n$.

The matrix denoted **A** in (20) has the negatives of the coefficients a_k from the left-hand side of (18) in its first column. This matrix has all 1's on the diagonal above the main diagonal. Except for the a_k's and the 1's, every other element of **A** is 0.

The vector denoted **b** in (21) is of dimension n, but it has exactly $m + 1$ nonzero elements. The elements of **b** are the coefficients b_k on the right-hand side of (18), starting with b_m on the bottom and working up as far as the coefficients go. Since we explicitly assume $m < n$ here, it will always fit.

Now consider the vector difference equation

$$\mathbf{X}(t + 1) = \mathbf{A}\mathbf{X}(t) + \mathbf{b}V(t) \tag{22}$$

If (22) is written out as a set of simultaneous scalar equations, we get

$$X_1(t+1) = -a_1 X_1(t) + X_2(t)$$

$$X_2(t+1) = -a_2 X_1(t) + X_3(t)$$

$$\vdots$$

$$X_{n-1}(t+1) = -a_{n-1} X_1(t) + X_n(t) + b_{m-1} V(t)$$

$$X_n(t+1) = -a_n X_1(t) + b_m V(t)$$

Increase the time index in the next-to-last equation by 1:

$$X_{n-1}(t+2) = -a_{n-1} X_1(t+1) + X_n(t+1) + b_{m-1} V(t+1)$$

Substitute the last equation on the right-hand side for $X_n(t+1)$:

$$X_{n-1}(t+2) = -a_{n-1} X_1(t+1) - a_n X_1(t) + b_{m-1} V(t+1) + b_m V(t)$$

Keep increasing the time index on successive equations, working up from the bottom and then substituting in this way. Eventually all of the equations are combined into the single equation

$$\begin{aligned} X_1(t+n) = {} & -a_1 X_1(t+n-1) - a_2 X_1(t+n-2) \\ & \cdots - a_{n-1} X_1(t+1) - a_n X_1(t) + b_0 V(t+m) \\ & + b_1 V(t+m-1) + \cdots + b_{m-1} V(t+1) + b_m V(t) \end{aligned} \tag{23}$$

Comparing (23) and (18), we see that they are the same if we take $X_1(t) = X(t)$. Therefore the vector equation (22) is equivalent to (18). We may solve (22) and then pick out the X_1 component, in order to find the solution to (18).

We may solve (22) exactly as in (2):

$$\mathbf{X}(1) = \mathbf{A}\mathbf{X}(0) + \mathbf{b}V(0)$$

$$\begin{aligned} \mathbf{X}(2) &= \mathbf{A}\mathbf{X}(1) + \mathbf{b}V(1) \\ &= \mathbf{A}^2\mathbf{X}(0) + \mathbf{A}\mathbf{b}V(0) + \mathbf{b}V(1) \end{aligned}$$

$$\begin{aligned} \mathbf{X}(3) &= \mathbf{A}\mathbf{X}(2) + \mathbf{b}V(2) \\ &= \mathbf{A}^3\mathbf{X}(0) + \mathbf{A}^2\mathbf{b}V(0) + \mathbf{A}\mathbf{b}V(1) + \mathbf{b}V(2) \end{aligned}$$

Again we see the general formula, which may be checked by direct substitution:

$$\mathbf{X}(t) = \mathbf{A}^t\mathbf{X}(0) + \sum_{k=0}^{t-1} \mathbf{A}^{t-k-1}\mathbf{b}V(k) \tag{24}$$

Since (18) was an nth-order equation, we need n initial conditions, say, $X(0)$, $X(1)$, $X(2), \ldots, X(n-1)$. Knowing these, we can compute the components of the initial state $X_1(0), X_2(0), \ldots, X_n(0)$.

The specific $\mathbf{A}$ and $\mathbf{b}$ given in (20) and (21) were only intended as an example to show that the nth-order scalar difference equation (18) could be converted to the first-order vector difference equation (22). However, the form (22) is much more general than that. Actually, $\mathbf{A}$ and $\mathbf{b}$ may be chosen almost arbitrarily. We see that for any $\mathbf{A}$ and $\mathbf{b}$, the solution to (22) is still given by (24).

Since $\mathbf{X}(t)$ is an n-dimensional vector whereas the input process $\{V(\cdot)\}$ is scalar valued, the question arises whether the vector $\mathbf{X}(t)$ can be expected to be found anywhere in the n-dimensional space R_n or whether it may be confined to some subspace. From (24), we see that if the vectors

$$\mathbf{b}, \mathbf{A}\mathbf{b}, \mathbf{A}^2\mathbf{b}, \ldots, \mathbf{A}^{n-1}\mathbf{b}$$

are all linearly independent, then the set of values $V(0), V(1), \ldots, V(n-1)$ are being mapped into $\mathbf{X}(n)$ by a nonsingular mapping, and therefore the distribution of $\mathbf{X}(n)$ will not be singular. We recognize this condition as the controllability criterion from control theory, which now shows up in a new context!

Calculation of Mean Vector and Covariance Matrix

Let us now specifically assume that $\{V(\cdot)\}$ is a WGN sequence as before and that $\mathbf{X}(0)$ is a Gaussian random vector, independent of $\{V(\cdot)\}$, with mean vector $\mathbf{m}$ and covariance matrix $\mathbf{C}_0$. Again we use the facts that the mean of a sum of random vectors is the sum of the means, and the covariance of a sum of independent random vectors is the sum of the covariances.

Make the definitions

$$E[\mathbf{X}(t)] = \boldsymbol{\mu}(t) \tag{25}$$

$$E\{[\mathbf{X}(t) - \boldsymbol{\mu}(t)][\mathbf{X}(s) - \mu(s)]^T\} = \mathbf{C}(t, s) \tag{26}$$

Since $E[V(t)] = 0$, from (24) we have at once

$$\boldsymbol{\mu}(t) = \mathbf{A}^t\mathbf{m} \tag{27}$$

Taking the expected value of both sides of (22) gives

$$\boldsymbol{\mu}(t+1) = \mathbf{A}\boldsymbol{\mu}(t) \tag{28}$$

It is obvious that (27) is the solution to (28), with the initial condition $\boldsymbol{\mu}(0) = \mathbf{m}$.

Now using (24) to calculate covariances, we have

$$\mathbf{C}(t, t) = \mathbf{A}^t\mathbf{C}_0(\mathbf{A}^t)^T + \sum_{k=0}^{t-1} (\mathbf{A}^{t-k-1}\mathbf{b})(\mathbf{A}^{t-k-1}\mathbf{b})^T \tag{29}$$

The question of finding $\mathbf{C}(t, s)$ for $t \neq s$, and also the issue of whether $\mathbf{C}(t, s)$ obeys a difference equation, requires us to proceed more stealthily. The first step is to subtract (28) from (22).

$$[\mathbf{X}(t+1) - \boldsymbol{\mu}(t+1)] = \mathbf{A}[\mathbf{X}(t) - \boldsymbol{\mu}(t)] + \mathbf{b}V(t) \tag{30}$$

Now multiply equation (30) on the right by the row vector $[\mathbf{X}(s) - \boldsymbol{\mu}(s)]^T$:

$$\begin{aligned}[\mathbf{X}(t+1) - \boldsymbol{\mu}(t+1)][\mathbf{X}(s) - \boldsymbol{\mu}(s)]^T \\ = \mathbf{A}[\mathbf{X}(t) - \boldsymbol{\mu}(t)][\mathbf{X}(s) - \boldsymbol{\mu}(s)]^T + V(t)\mathbf{b}[X(s) - \boldsymbol{\mu}(s)]^T\end{aligned} \tag{31}$$

Assume $t \geqslant s$. In that case $V(t)$ is independent of $[\mathbf{X}(s) - \boldsymbol{\mu}(s)]$. Take the expected value of (31) for that case only:

$$\mathbf{C}(t+1, s) = \mathbf{A}C(t, s), \qquad t \geq s \tag{32}$$

Obviously the solution of this difference equation in the first variable, keeping the second variable constant, is

$$\mathbf{C}(t, s) = \mathbf{A}^{t-s}\mathbf{C}(s, s), \qquad t \geq s \tag{33}$$

In order to find the result for $t \leq s$, it is necessary to repeat the entire procedure from (30), reversing the roles of t and s. Either doing this explicitly, or arguing from symmetry, we find

$$\mathbf{C}(t, s) = \mathbf{C}(t, t)(\mathbf{A}^{s-t})^T, \qquad t \leq s \tag{34}$$

Thus we have obtained expressions for $\mathbf{C}(t, s)$ for all values of t and s. When both arguments are equal, we use (29), which is valid for all $t > 0$. When the arguments are unequal, we use (29) first, followed by (33) or (34) as appropriate.

The one question remaining is whether $\mathbf{C}(t, t)$ itself obeys some sort of difference equation. Let us examine that now. The equations we have to work with are just (30), which governs the evolution of the $\{\mathbf{X}(t)\}$ process, and (26), which defines $\mathbf{C}(t, s)$. From (26) we have

$$\mathbf{C}(t+1, t+1) = E\{[\mathbf{X}(t+1) - \boldsymbol{\mu}(t+1)][\mathbf{X}(t+1) - \boldsymbol{\mu}(t+1)]^T\} \tag{35}$$

By definition of the WGN sequence $\{V(t)\}$, $V(t)$ is independent of $\mathbf{X}(t)$. Therefore, multiplying (30) on the right by its own transpose and taking expected value yields

$$\mathbf{C}(t+1, t+1) = \mathbf{A}\mathbf{C}(t, t)\mathbf{A}^T + \mathbf{b}\mathbf{b}^T \tag{36}$$

where we have assumed

$$E[V^2(t)] = 1 \tag{37}$$

for each t.

We leave it as one of the problems to verify that the expression (29) we obtained previously is the solution to (36).

So far, we have used the difference equation (1) as a way of generating a correlated random sequence from a WGN sequence. The resulting sequence was used to illustrate the Markov property. We then generalized this idea to (18). The study of (18) was facilitated by introduction of vectors and matrices; this procedure is known as the *state-space technique*. In that form, (18) was converted to (22). The solution to (22) was given by (24), from which we obtained (27) as the explicit expression for the mean of the $\{\mathbf{X}(t)\}$ process at any time and (29), (33), and (34) as explicit expressions for the autocovariance of the $\{X(t)\}$ process.

Even though all of the parameters in (22), that is, the matrix $\mathbf{A}$ and the vector $\mathbf{b}$, were constant, nevertheless the autocovariance $\mathbf{C}(t, s)$ turns out not to depend only on the difference $t - s$, but actually depends on t and s separately. However, we do see from (33) and (34) that if $\mathbf{C}(s, s)$ and $\mathbf{C}(t, t)$ were constant, that is, did not vary with s or t, then in fact $\mathbf{C}(t, s)$ *would* depend only on the difference $t - s$. The question is, therefore, whether the expression (29) converges to a limit as $t \to \infty$.

It turns out that whenever $\mathbf{A}$ is stable, that is, all of its eigenvalues have magnitude less than one, then (29) does in fact converge. This can be shown, tediously, by transforming $\mathbf{A}$ to Jordan canonical form and then examining the summation element by element. The resulting covariance matrix, which we denote $\mathbf{C}_\infty$, must therefore be a solution to the algebraic equation that results when we substitute

$$\mathbf{C}_\infty = \mathbf{C}(t,t) = \mathbf{C}(t+1, t+1) \tag{38}$$

into (36), specifically

$$\mathbf{C}_\infty = \mathbf{A}\mathbf{C}_\infty\mathbf{A}^T + \mathbf{b}\mathbf{b}^T \tag{39}$$

As noted in the discussion following (24), $\mathbf{C}_\infty$ will be nonsingular if and only if the pair $(\mathbf{A}, \mathbf{b})$ is controllable in the sense of control theory.

Thus, in general, given a time-invariant linear system excited by stationary white noise, starting from a specified initial condition at $t = 0$, we have to wait a long time (theoretically infinite; in practice, several times the longest time constant of the system) for the output process to become stationary. This is the stochastic equivalent of the familiar deterministic phenomenon of a "starting transient." The only way to obtain a stationary process from the beginning would be if we could specify the distribution of the initial state $\mathbf{X}(0)$.

We have said that $\mathbf{X}(0)$ is Gaussian with mean $\mathbf{m}$ and covariance $\mathbf{C}_0$. The only way to get a stationary process immediately would be to be able to specify

$$\begin{aligned} \mathbf{m} &= 0 \\ \mathbf{C}_0 &= \mathbf{C}_\infty \end{aligned} \tag{40}$$

Time-Varying State Space Models

Since in general we will have to cope with a nonstationary process whenever we consider a state-space model with arbitrary initial conditions starting at a specific finite time (i.e., not having started in the infinite past, as did the processes considered in Chapters 5 and 6), then we may as well generalize our basic model to allow time-varying parameters. Also, let us generalize from a scalar white noise input sequence to a vector sequence.

By a vector WGN sequence, we mean a sequence of Gaussian random vectors $\mathbf{V}(t)$, $t = 0, 1, 2, \ldots$, such that $\mathbf{V}(t)$ and $\mathbf{V}(s)$ are independent whenever $t \neq s$. At a given time t, however, the components of the vector

$\mathbf{V}(t)$ need not be independent of one another. We assume that the mean of $\mathbf{V}(t)$ is zero for every t and that its covariance is $\mathbf{Q}(t)$:

$$E\mathbf{V}(t)\mathbf{V}^T(t) = \mathbf{Q}(t) \tag{41}$$

Let us assume that $\mathbf{V}(t)$ is of dimension m, so $\mathbf{Q}(t)$ is $m \times m$.

We now generalize (22) to

$$\mathbf{X}(t+1) = \mathbf{A}(t)\mathbf{X}(t) + \mathbf{B}(t)\mathbf{V}(t) \tag{42}$$

where $\mathbf{X}(0)$ is random with mean $\mathbf{m}$ and covariance $\mathbf{C}_0$ as before, and $\mathbf{X}(0)$ is independent of the entire sequence $\{\mathbf{V}(t)\}$. As before, we assume $\mathbf{X}(t)$ is of dimension n, so $\mathbf{A}(t)$ is $n \times n$ and $\mathbf{B}(t)$ is $n \times m$.

Define the transition matrix $\mathbf{\Phi}(t, s)$ by

$$\mathbf{\Phi}(t, s) = \prod_{k=s}^{t-1} \mathbf{A}(k) \tag{43}$$

for $t > s$. The formula (24) for the solution to (22) generalizes to the following formula for the solution of (42):

$$\mathbf{X}(t) = \mathbf{\Phi}(t,0)\mathbf{X}(0) + \sum_{k=0}^{t-1} \mathbf{\Phi}(t, k+1)\mathbf{B}(k)\mathbf{V}(k) \tag{44}$$

The definitions (25) and (26) continue to apply. It is straightforward to show that (27) now becomes

$$\boldsymbol{\mu}(t) = \mathbf{\Phi}(t,0)\mathbf{m} \tag{45}$$

By following the same procedure as before, one can derive the formulas for $\mathbf{C}(t, s)$:

$$\begin{aligned} \mathbf{C}(t, s) &= \mathbf{\Phi}(t, s)\mathbf{C}(s, s), \qquad t \geq s \\ \mathbf{C}(t, s) &= \mathbf{C}(t, t)\mathbf{\Phi}^T(s, t), \qquad t \leq s \end{aligned} \tag{46}$$

where $\mathbf{C}(t, t)$ is the solution to the matrix difference equation

$$\mathbf{C}(t+1, t+1) = \mathbf{A}(t)\mathbf{C}(t, t)\mathbf{A}^T(t) + \mathbf{B}(t)\mathbf{Q}(t)\mathbf{B}^T(t) \tag{47}$$

By standard methods, one obtains the following solution to (47),

$$\mathbf{C}(t,t) = \mathbf{\Phi}(t,0)\mathbf{C}_0\mathbf{\Phi}^T(t,0) + \sum_{K=0}^{t-1} \mathbf{\Phi}(t,k+1)\mathbf{B}(k)\mathbf{Q}(k)\mathbf{B}^T(k)\mathbf{\Phi}^T(t,k+1) \tag{48}$$

or one can simply take the covariance of (44).

The time-varying vector stochastic difference equation (42) is widely used as a model in many current applications of stochastic process theory, wherever a linear state-space model seems appropriate. In Control Theory, this equation is generalized to include an additional control input term. In Communication Theory, it is used for modeling signal processes when a nonstationary model is required. The actual signal is often obtained as a projection of the state:

$$\mathbf{S}(t) = \mathbf{H}(t)\mathbf{X}(t) \tag{49}$$

where $\mathbf{H}(t)$ is a known time-varying matrix of dimension $r \times n$ and $\mathbf{S}(t)$ is an r-dimensional vector (typically $r < n$) representing the signal process.

In many applications, a noisy measurement device or transmission of the signal over a noisy channel is represented by more additive white noise, so the observed output or received signal is $\mathbf{S}(t)$ plus noise.

Problems

1. Refer to $\mathbf{C}(t,t)$ as given by (29). Considering this as a matrix-valued function of the single-integer-valued argument t, show that it is the unique solution to the first-order difference equation (36), with initial condition $\mathbf{C}(0,0) = \mathbf{C}_0$.
2. Given the scalar difference equation $Y(t+3) - 6Y(t+2) + 11Y(t+1) - 6Y(t) = W(t)$ with nonrandom initial conditions $Y(1) = 6$, $Y(2) = 10$, and $Y(3) = 24$, let $\{W(t),\ t = 1,2,\dots\}$ be a sequence of independent random variables, each with mean zero and unit variance.
   a. Compute $\mu(t) = E\{Y(t)\}$.
   b. Compute $c(t,s) = E\{[Y(t) - \mu(t)][Y(s) - \mu(s)]\}$.
3. Given the discrete-time stochastic linear system

$$\mathbf{X}(t+1) = \mathbf{F}\mathbf{X}(t) + \mathbf{W}(t)$$

where the initial state $\mathbf{X}(0)$ is random with mean zero and covariance $\mathbf{\Lambda}$.

If $\{W(t)\}$ is a WGN sequence independent of $\mathbf{X}(0)$, with zero mean and

$$E[W(t)W^T(t)] = \begin{bmatrix} 2 & 1 \\ 1 & 2 \end{bmatrix}$$

and if

$$\mathbf{F} = \begin{bmatrix} 0 & 1 \\ -0.5 & 1 \end{bmatrix}$$

find the value that $\boldsymbol{\Lambda}$ must be in order for $\{X(t)\}$ to be stationary from the beginning.

4. Let $\{W(\cdot)\}$ be scalar discrete-time WGN, with mean zero and unit variance. Consider the vector difference equation

$$\begin{bmatrix} X_1(t+1) \\ X_2(t+1) \end{bmatrix} = \begin{bmatrix} 0 & 1 \\ -0.24 & 1 \end{bmatrix} \begin{bmatrix} X_1(t) \\ X_2(t) \end{bmatrix} + \begin{bmatrix} 0 \\ 1 \end{bmatrix} W(t)$$

for $t = 0, 1, 2, \ldots$. Let the initial state $\begin{bmatrix} X_1(0) \\ X_2(0) \end{bmatrix}$ be random, independent of $\{W(\cdot)\}$. Define

$$\boldsymbol{\mu} = E\left\{\begin{bmatrix} X_1(0) \\ X_2(0) \end{bmatrix}\right\}$$

$$\boldsymbol{\Lambda} = E\left\{\left[\begin{bmatrix} X_1(0) \\ X_2(0) \end{bmatrix} - \boldsymbol{\mu}\right]\left[\begin{bmatrix} X_1(0) \\ X_2(0) \end{bmatrix} - \boldsymbol{\mu}\right]^T\right\}$$

How should $\boldsymbol{\mu}$ and $\boldsymbol{\Lambda}$ be chosen so that the process $\{\mathbf{X}(t)\}$ is stationary from the beginning?

5. Given the vector difference equation

$$\mathbf{X}(t+1) = \mathbf{A}\mathbf{X}(t) + \mathbf{V}(t), \qquad t = 0, 1, 2, \ldots$$

with

$$\mathbf{A} = \begin{bmatrix} \cos\theta & \sin\theta \\ -\sin\theta & \cos\theta \end{bmatrix}$$

where θ is a parameter, $-\pi \le \theta \le \pi$. Let $\mathbf{V}(t)$ be vector WGN with zero mean and covariance $2\mathbf{I}$ for every t. Let the initial state $\mathbf{X}(0)$ be independent of $\{W(\cdot)\}$, with mean zero and covariance $10\mathbf{I}$. Find $\mathbf{C}(t, t)$, the matrix defined in (26) with $s = t$, for every t.

6. Let $\{X(t),\ t = 0, 1, 2, \dots\}$ be a discrete-time scalar *Gaussian* process. In a manner analogous to (14) of the text, define the conditional probability density $p(x(t+1)|x(t), \dots, x(0))$. Prove that if it holds that

$$E\{X(t+1)|X(t),\ X(t-1), \dots, X(0)\} = E\{X(t+1)|X(t)\}$$

then also

$$p(x(t+1)|x(t), x(t-1), \dots, x(0)) = p(x(t+1)|x(t))$$

Is this true if $\{X(\cdot)\}$ is not Gaussian?

Chapter 11

Discrete-Time Kalman Filtering

Introduction

The results we present in this chapter are somewhat limited in terms of the full scope of information that exists on the Kalman filter. There are three reasons for that. First, this will be the initial acquaintance with the Kalman filter for many students. Experience has shown that information overload occurs rather readily with this subject. It is better to provide a preliminary introduction to the concept of the Kalman filter (i.e., what it is, what it does) than it is to try to provide the most general form of the equations or to proceed too quickly to a discussion of implementation on the computer. Most students need time to digest and assimilate the introduction before proceeding to the details of implementation. Failure to allow for this seems to leave the student with a "gee whiz" attitude toward the filter. He or she has a set of equations, but is not confident of the theoretical premises on which they are based. The student will therefore tend to treat the filter as a "magic box," which is somehow the panacea for all filtering and estimation problems, without being aware of the limitations.

The second reason for restricting the scope of this chapter is that, consistent with the philosophy just stated, students will have other courses beyond this one in which the Kalman filter will again be presented and many of its ramifications will be discussed.

Third, the author's reason for including this subject at all in this book is not so much to teach the Kalman filter as an end in itself as it is to use this topic as a vehicle for bringing students to a whole set of concepts they may have met before, but which are worth examining again in a specifically stochastic context. These concepts are: causality, invertibility, orthogonal decomposition, and recursivity. Rather than present these concepts by means of abstract definitions, or by means of theorems and proofs, in this chapter the algorithm from Appendix 1 is used to provide a concrete algebraic illustration of the ideas.

The user of a Kalman filter usually has a specific problem he is trying to solve. He wishes to know the sequence of states of a dynamic system which he cannot observe directly. What he does have is a data sequence that carries some information about the states he wishes to know. In lieu of the actual sequence of states, our user is willing to accept a sequence of *estimates* of the states, based on the data to which he does have access, provided the estimates are suitably accurate for his purpose.

The Kalman filter converts the sequence of data into the sequence of estimates, in real time. That is, at each point in time, the filter provides an estimate of the current state of the system, based on all of the observations that have been made so far. Moreover, it also provides a theoretical measure of the error associated with each estimate.

The algorithm in Appendix 1 computes the inverse of a covariance matrix by successively inverting a sequence of submatrices. It simultaneously provides the triangular factorization of the covariance matrix by successively computing the triangular factors of the submatrices.

We will use the lower triangular form as an explicit model for a causal operator. We will show how the structure of the algorithm automatically tests the invertibility of the covariance matrix, and how our causal operators are always also causally invertible. We will show how our triangular factorization automatically performs Gram–Schmidt orthogonalization of the data sequence to yield a sequence of orthogonal r.v.'s which carries the same information. Our discussion of the Hilbert space of second-order r.v.'s in Chapter 1 is specifically relevant here. Finally, we will see what it is about the signal model adopted in Kalman filtering theory which permits one to obtain a recursive filter having a fixed, finite dimension.

Problem Formulation

Let us now proceed to justify these remarks by introducing equations and deriving specific results. The signal model developed in Chapter 10 is the one we shall use here, namely,

$$\mathbf{X}(t+1) = \mathbf{A}(t)\mathbf{X}(t) + \mathbf{B}(t)\mathbf{V}(t) \tag{1}$$

Here, $\{\mathbf{V}(t)\}$ is WGN with zero mean and covariance

$$E[\mathbf{V}(t)\mathbf{V}^T(t)] = \mathbf{Q}(t) \tag{2}$$

The initial state $\mathbf{X}(0)$ is Gaussian with mean $\mathbf{0}$ and covariance $\mathbf{P}_0$, and is independent of $\{\mathbf{V}(t)\}$.

One of the ways the scope of this discussion will be restricted is that our signal and data models will be scalar valued. This is so we can exploit the procedure based on Appendix 1 with a minimum of notational problems.

Let $\mathbf{h}(t)$ be a known time-varying vector of dimension n. The signal output from our model is

$$S(t) = \mathbf{h}^T(t)\mathbf{X}(t) \tag{3}$$

The actual data that serves as fodder for the filter is this signal sequence corrupted with added noise:

$$Y(t) = S(t) + N(t) \tag{4}$$

Here $\{N(t)\}$ is a zero mean scalar WGN sequence with covariance

$$E[N^2(t)] = R(t) > 0 \tag{5}$$

$\mathbf{X}$, $\{\mathbf{V}(t)\}$, and $\{N(t)\}$ are mutually independent.

We assume that our Kalman filter user receives the sequence $Y(1), Y(2), \ldots$. He wants to estimate the sequence of states $\mathbf{X}(1), \mathbf{X}(2), \ldots$. Specifically, at each time t he wants to construct an estimate $\hat{\mathbf{X}}(t)$ of the current state $\mathbf{X}(t)$, based on the data $Y(1), Y(2), \ldots, Y(t)$.

Our first step is to define a vector $\mathbf{Z}(t)$ of variable dimension.

$$\mathbf{Z}(t) = \begin{bmatrix} Y(1) \\ Y(2) \\ \vdots \\ Y(t) \end{bmatrix} \tag{6}$$

Now recall the brief discussion of statistical estimation given at the end of Chapter 3, where it was pointed out that in order to talk about an optimal estimate it is necessary to specify a loss function. We adopt as our criterion the mean squared error $E\{\|\mathbf{X}(t) - \hat{\mathbf{X}}(t)\|^2\}$. Thus, we really have not just one estimation problem but a whole sequence of them, with a corresponding loss function specified separately at each time t. Of course, we are anticipating that there will be so much structure in the solution as a consequence of the model we have adopted, that there will be a strong relationship between the estimates at successive times.

At each time t we want to find the best estimate of $\mathbf{X}(t)$ in the sense of this criterion, based on the available vector $\mathbf{Z}(t)$. Since as a consequence of our assumptions there is a known joint distribution of the pair $\mathbf{X}(t)$ and $\mathbf{Z}(t)$, we are dealing with a Bayesian model. Therefore, we are seeking the minimum mean squared error Bayesian estimate, which we saw in Chapter 3 is provided by the conditional mean:

$$\hat{\mathbf{X}}(t) = E\{\mathbf{X}(t)|\mathbf{Z}(t)\} \tag{7}$$

Our task now is, given the model specified by equations (1)–(6), compute (7) for each t. Clearly, from our assumptions, at time t the vector $\mathbf{Z}(t)$ is a Gaussian random vector, with zero mean, of dimension t. The covariance matrix to which we will be applying the results of Appendix 1 is the covariance of $\mathbf{Z}(t)$. Therefore, we adopt a notation that facilitates comparison with Appendix 1:

$$E\{\mathbf{Z}(t)\mathbf{Z}^T(t)\} = \mathbf{C}_t \tag{8}$$

LDL^T Factorization, the Innovations Sequence, and the Update Formula

Evidently $\mathbf{C}_t$ is a positive definite symmetric matrix of dimension t. As shown in Appendix 1, for each t it can be factored as

$$\mathbf{C}_t = \mathbf{L}_t\mathbf{D}_t\mathbf{L}_t^T \tag{9}$$

where $\mathbf{L}_t$ is a lower triangular matrix with all 1's on the main diagonal, and $\mathbf{D}_t$ is a diagonal matrix with all positive elements on the main diagonal. The diagonal elements of D_t are calculated sequentially: $d_1, d_2, \ldots, d_t$. The proof in Appendix 1 shows that, so long as C_t is positive definite, these elements will all be positive. If C_t should lose positive definiteness as we go from t to $t + 1$, then d_{t+1} will fail to be positive. Therefore, at each step we automatically test the invertibility of C_t.

Since $\mathbf{L}_t$ is necessarily nonsingular, it always has an inverse, $\mathbf{L}_t^{-1}$. Let us introduce another vector that we designate by a lowercase Greek letter even though it is a random vector:

$$\zeta(t) = \mathbf{L}_t^{-1}\mathbf{Z}(t) \tag{10}$$

We are deviating from the convention of denoting random quantities by capital letters here and in the next equation in order to use notation that has become somewhat customary. Denote the individual components of

$\zeta(t)$ by the scalars $\nu(1), \nu(2), \ldots, \nu(t)$:

$$\zeta(t) = \begin{bmatrix} \nu(1) \\ \nu(2) \\ \vdots \\ \nu(t) \end{bmatrix} \tag{11}$$

Now, if we know the sequence of matrices $\{\mathbf{L}_t^{-1}\}$, then as the data sequence $Y(1), Y(2), \ldots, Y(t)$ comes in, we can form the vector $\mathbf{Z}(t)$ according to (6), compute $\zeta(t)$ by (10), and then construct the sequence $\nu(1), \nu(2), \ldots, \nu(t)$ by (11). The first thing we have to show is that our notation is well chosen, in the sense that for every t, the *previously computed* ν's *do not change*. For example, we need to be assured that $\nu(1)$, $\nu(2)$, and $\nu(3)$ computed in the way that we just described from $\mathbf{Z}(4)$ are the same as $\nu(1)$, $\nu(2)$, and $\nu(3)$ computed from, say, $\mathbf{Z}(7)$. Evidently that won't happen automatically unless the sequence of matrices $\{L_t^{-1}\}$ enjoys some special property.

The required property is guaranteed by equation (A1.6) of Appendix 1. As we go from t to $t+1$, the upper left $t \times t$ submatrix of $\mathbf{L}_{t+1}^{-1}$ is precisely $\mathbf{L}_t^{-1}$. All we do is put another 1 on the last diagonal spot and fill out the bottom row appropriately, in order to go from $\mathbf{L}_t^{-1}$ to $\mathbf{L}_{t+1}^{-1}$. Thus we have

$$\begin{bmatrix} \nu(1) \\ \nu(2) \\ \vdots \\ \nu(t) \\ \nu(t+1) \end{bmatrix} = \begin{bmatrix} \mathbf{L}_t^{-1} & \vdots & \mathbf{0} \\ \cdots & \cdots & \cdots \\ \text{stuff} & \vdots & 1 \end{bmatrix} \begin{bmatrix} Y(1) \\ Y(2) \\ \vdots \\ Y(t) \\ Y(t+1) \end{bmatrix} \tag{12}$$

Basically, it is the fact that the inverse of a lower triangular matrix is itself lower triangular which allows this construction to work. Therefore, we have constructed a well-defined, unique new random sequence $\nu(1), \nu(2), \ldots, \nu(t)$ from the given data sequence $Y(1), Y(2), \ldots, Y(t)$. The sequence $\{\nu(t)\}$ is called the *innovations sequence*. The reason for using this name will become clear later.

From equations (8)–(10), we see that we must have

$$E\{\zeta(t)\zeta^T(t)\} = \mathbf{D}_t \tag{13}$$

Since $\mathbf{D}_t$ is a diagonal matrix, (13) means that the components of $\zeta(t)$ are mutually orthogonal. Since $\zeta(t)$ is zero mean Gaussian, it means that the members of the sequence $\{\nu(t)\}$ are mutually independent, that is, $\{\nu(t)\}$ is a WGN sequence with time-varying covariance.

Now define the following cross-covariance matrices:

$$\mathbf{C}_{XZ}(t) = E\{\mathbf{X}(t)\mathbf{Z}^T(t)\} \tag{14}$$

$$\mathbf{C}_{X\zeta}(t) = E\{\mathbf{X}(t)\boldsymbol{\zeta}^T(t)\} \tag{15}$$

In equation (44) of Chapter 3, we gave the following equation for $E\{\mathbf{X}|\mathbf{Y}\}$ when $\mathbf{X}$ is a Gaussian random vector with mean $\boldsymbol{\mu}_X$, $\mathbf{Y}$ is a Gaussian random vector with mean $\boldsymbol{\mu}_Y$, and also

$$E\left\{(\mathbf{X} - \boldsymbol{\mu}_X)(\mathbf{Y} - \boldsymbol{\mu}_Y)^T\right\} = \mathbf{C}_{XY}$$

$$E\left\{(\mathbf{Y} - \boldsymbol{\mu}_Y)(\mathbf{Y} - \boldsymbol{\mu}_Y)^T\right\} = \mathbf{C}_{YY}$$

Then

$$E\{\mathbf{X}|\mathbf{Y}\} = \boldsymbol{\mu}_X + \mathbf{C}_{XY}\mathbf{C}_{YY}^{-1}(\mathbf{Y} - \boldsymbol{\mu}_Y) \tag{16}$$

In the present case all means are zero. Recalling equations (8) and (13)–(15), we see that this formula states that

$$E\{\mathbf{X}(t)|\mathbf{Z}(t)\} = \mathbf{C}_{XZ}(t)\mathbf{C}_t^{-1}\mathbf{Z}(t) \tag{17}$$

$$E\{\mathbf{X}(t)|\boldsymbol{\zeta}(t)\} = \mathbf{C}_{X\zeta}(t)\mathbf{D}_t^{-1}\boldsymbol{\zeta}(t) \tag{18}$$

Now since $\mathbf{Z}(t)$ and $\boldsymbol{\zeta}(t)$ are related by a nonsingular transformation, there can be no statistical information lost if we base our estimate of $\mathbf{X}(t)$ on $\boldsymbol{\zeta}(t)$ rather than $\mathbf{Z}(t)$. That is, even though the expressions are different, both (17) and (18) must be the *same estimate*, namely $\hat{\mathbf{X}}(t)$. All other considerations being equal, it seems (18) would be preferable to use because it is much easier to invert a diagonal matrix than an arbitrary covariance matrix. In fact, since we start with $\mathbf{C}_t$ and not $\mathbf{D}_t$, we have to use the algorithm of Appendix 1 to find $\mathbf{D}_t$, so it comes to the same thing. However, the derivation of the Kalman filter is expedited by using (18).

Our immediate goal is to see how much we can say about the *structure* of the filter just on the basis of (18) and Appendix 1, *without* bringing in all the equations that describe our model and obtaining explicit filter equations. Presently, we will get the explicit equations.

In terms of structure alone, we are going to give another reason for preferring (18) to (17). Consider that, by (15),

$$\mathbf{C}_{X\zeta}(t) = E\{\mathbf{X}(t)[\nu(1)\nu(2)\cdots\nu(t)]\} \tag{19}$$

This says that the kth column of $\mathbf{C}_{X\zeta}(t)$ is simply $E\{\mathbf{X}(t)\nu(k)\}$.

Let us now adopt a notation that brings these column vectors explicitly into our expressions. Assume for the moment that t is fixed at some particular positive integer. Define for each integer k, $1 \leq k \leq t$, the vector

$$\boldsymbol{\alpha}_{tk} = E\{\mathbf{X}(t)\nu(k)\} \tag{20}$$

The diagonal elements of $\mathbf{D}_t$ are denoted by $d_1, d_2, \ldots, d_t$, respectively. Thus, recalling (11), we may rewrite (18) as

$$E\{\mathbf{X}(t)|\zeta(t)\} = \sum_{k=1}^{t} \boldsymbol{\alpha}_{tk}\frac{\nu(k)}{d_k} \tag{21}$$

Going back to our fundamental estimation formula (16) and making the appropriate definitions, it is easy to verify that if we wanted to compute the estimate of $\mathbf{X}(t)$ based on $\zeta(t-1)$ rather than $\zeta(t)$, all that happens is that we delete the last term from (21):

$$E\{\mathbf{X}(t)|\zeta(t-1)\} = \sum_{k=1}^{t-1} \boldsymbol{\alpha}_{tk}\frac{\nu(k)}{d_k} \tag{22}$$

We may look at this another way. Suppose we already had computed $E\{\mathbf{X}(t)|\zeta(t-1)\}$. When we receive the additional data value $\nu(t)$, all we have to do is add one term:

$$E\{\mathbf{X}(t)|\zeta(t)\} = E\{\mathbf{X}(t)|\zeta(t-1)\} + \boldsymbol{\alpha}_{tt}\frac{\nu(t)}{d_t} \tag{23}$$

Contrast the simple situation in (23) with what happens if we work with the estimation formula (17). Analogously to (19) we may write

$$\mathbf{C}_{XZ}(t) = E\{\mathbf{X}(t)[Y(1)Y(2)\cdots\ Y(t)]\} \tag{24}$$

Analogously to (20) we may define

$$\boldsymbol{\beta}_k = E\{\mathbf{X}(t)Y(k)\} \tag{25}$$

Now define the t-dimensional vector

$$\boldsymbol{\phi} = [\phi_1\phi_2\ \cdots\ \phi_t]^T = \mathbf{C}_t^{-1}\mathbf{Z}(t) \tag{26}$$

Then we may write

$$E\{\mathbf{X}(t)|\mathbf{Z}(t)\} = \sum_{k=1}^{t} \boldsymbol{\beta}_k\phi_k \tag{27}$$

Also define the $(t-1)$-dimensional vector

$$\boldsymbol{\psi} = [\psi_1 \psi_2 \cdots \psi_{t-1}] = \mathbf{C}_{t-1}^{-1}\mathbf{Z}(t-1) \tag{28}$$

It is easy to verify that

$$E\{\mathbf{X}(t)|\mathbf{Z}(t-1)\} = \sum_{k=1}^{t-1} \boldsymbol{\beta}_k \psi_k \tag{29}$$

However, since $\mathbf{C}$ is not a diagonal covariance matrix, it is not true that the upper left $(t-1)$-dimensional submatrix of $\mathbf{C}_t^{-1}$ is $\mathbf{C}_{t-1}^{-1}$. This is in contrast to the relation between $\mathbf{D}_t^{-1}$ and $\mathbf{D}_{t-1}^{-1}$. The consequence of this is that the components of the vector $\boldsymbol{\psi}$ defined in (28) are not just the first $t-1$ components of $\boldsymbol{\phi}$ defined in (26), but in general, $\boldsymbol{\psi}$ and $\boldsymbol{\phi}$ are entirely different. Therefore, there is no simple formula relating $E\{\mathbf{X}(t)|\mathbf{Z}(t-1)\}$ and $E\{\mathbf{X}(t)|\mathbf{Z}(t)\}$ which corresponds to (23).

Equation (23) is extremely important in our derivation of the Kalman filter. It is called the *update* formula. It shows how to update the estimate of $\mathbf{X}(t)$ given $\zeta(t-1)$ when one more data point, $\nu(t)$, arrives, in order to obtain the estimate of the *same* quantity, $\mathbf{X}(t)$, given $\zeta(t)$.

We have just seen the importance of working with a statistically orthogonal sequence $\{\nu(t)\}$ rather than the original data sequence $\{Y(t)\}$ in order to obtain a simple update formula. This result is one of the key elements in obtaining a *recursive* form of the filter. *Recursive* means that we do not have to store the entire data vector $\mathbf{Z}(t)$, which continues to grow as t advances, and entirely recompute the estimate $\hat{\mathbf{X}}(t)$ at each stage. In a recursive filter, $\hat{\mathbf{X}}(t)$ itself serves as a summary of all the past data. It is only necessary to modify the estimate as we go along by suitably incorporating the new data as it arrives.

This is the time to discuss another very important property of the sequence $\{\nu(t)\}$. This sequence is not the only statistically orthogonal sequence that can be obtained from $\{Y(t)\}$ by an arbitrary linear transformation. For example, in (8) we defined $\mathbf{C}_t$ as the covariance of the data vector $\mathbf{Z}(t)$. Now, it is always possible in principle to compute the eigenvalues and eigenvectors of $\mathbf{C}_t$, and from that to find the modal matrix and the orthogonal transformation that diagonalizes $\mathbf{C}_t$. Let this diagonal representation be denoted as follows:

$$\mathbf{C}_t = \mathbf{O}_t \boldsymbol{\Lambda}_t \mathbf{O}_t^T \tag{30}$$

Here, $\mathbf{O}_t$ is an orthogonal matrix whose columns are the normalized eigenvectors of $\mathbf{C}_t$, and $\boldsymbol{\Lambda}_t$ is a diagonal matrix having the appropriate eigenvalues of $\mathbf{C}_t$ on the main diagonal.

Suppose now we define the vector $\mathbf{U}(t)$ by

$$\mathbf{U}(t) = \mathbf{O}_t^T \mathbf{Z}(t) \tag{31}$$

From (8) and (30) it follows that

$$E\{\mathbf{U}(t)\mathbf{U}^T(t)\} = \mathbf{\Lambda}_t \tag{32}$$

Equations (31) and (32) should be compared to equations (10) and (13), respectively. This shows that there is more than one transformation that will convert the vector $\mathbf{Z}(t)$ into a new vector having a *diagonal* covariance.

The point we want to emphasize here is that the special feature of the transformation in (10) is its *causality*. Keep in mind equations (6) and (11) and the fact that $\mathbf{L}_t^{-1}$ is *lower triangular*. Then, (10) says that $\nu(1)$ is computed from $Y(1)$ alone, $\nu(2)$ is computed from $Y(1)$ and $Y(2)$, $V(3)$ is computed from $Y(1)$, $Y(2)$, and $Y(3)$, and in general, $\nu(k)$ is computed from $Y(1), Y(2), \ldots, Y(k)$.

If this point is not clear, the reader should write out (10) elementwise until the above statement becomes evident. It is essential to grasp this point, because it is the connection between lower triangular matrices and causality. It is the basic reason that the author considers the algorithm in Appendix 1 so pertinent for gaining insight into the Kalman filter.

The difference between equations (9) and (30) is that, in equation (9), the covariance $\mathbf{C}_t$ has been diagonalized by a *lower triangular* (causal) transformation. The orthogonal matrix $\mathbf{O}_t$ in (30) in general will not be lower triangular. Therefore, in order to obtain one of the intermediate components of the vector $\mathbf{U}(t)$ in (31), we would need the entire vector $\mathbf{Z}(t)$.

The orthogonality of the sequence $\{\nu(t)\}$ which led to the simple formula (23) is only part of the qualities of $\{\nu(t)\}$ which permit achieving a recursive filter. The fact that an orthogonal decomposition permits one to compute one new term in a finite sum approximation without changing all the previous terms is an idea that should be familiar to the student from the study of Fourier series. We have just seen that there is more than one orthogonal decomposition possible. The feature that makes $\{\nu(t)\}$ special is that it can be obtained *causally* from $\{Y(t)\}$. Moreover, no information is lost by this procedure. We know this because the transformation is invertible, in fact, *causally invertible*.

Another way of looking at this, which is sometimes mentioned in the literature, is to view the members of $\{Y(t)\}$ as elements of the Hilbert space of second-order r.v.'s discussed in Chapter 1. The sequence $\{\nu(t)\}$ is generated by performing the Gram–Schmidt orthogonalization procedure on $\{Y(t)\}$. Since Gram–Schmidt is necessarily causal, this yields a causal transformation.

Once again: A key element in the recursivity of the Kalman filter is that the sequence $\{\nu(t)\}$ is a statistically orthogonal sequence that is derived causally and invertibly from the data sequence $\{Y(t)\}$.

Equation (23) shows us how to get from $E\{\mathbf{X}(t)|\boldsymbol{\zeta}(t-1)\}$ to $E\{\mathbf{X}(t)|\boldsymbol{\zeta}(t)\}$. Now, our objective is to obtain a filter that tracks the *current* state of the model. We now need to examine how to modify the estimate as the state changes from $\mathbf{X}(t)$ to $\mathbf{X}(t+1)$ without changing the data vector. That is, how do we get from $E\{\mathbf{X}(t)|\boldsymbol{\zeta}(t)\}$ to $E\{\mathbf{X}(t+1)|\boldsymbol{\zeta}(t)\}$? The formula that connects these two quantities is called the *propagate* formula because it shows how to propagate the state from $\mathbf{X}(t)$ to $\mathbf{X}(t+1)$. The quantity $E\{\mathbf{X}(t+1)|\boldsymbol{\zeta}(t)\}$ is sometimes called the *one-step prediction* because it is estimating the state one step ahead of the current state.

The Markov Model and the Propagate Formula

As we have just seen, we obtained the update formula (23) without making use of our model, equations (1)–(5). All we needed was the sequential nature of the data itself, represented by (6), and the results of the algorithm for sequential triangular factorization in Appendix 1. This is half the story for obtaining a recursive filter. The other half of the story involves the propagate formula, and this is where the model (1) comes in. If we did not have a vector Markov model, we could not get a propagate formula that would lead to a recursive filter.

Another way of saying this, which is the more prevalent viewpoint in communications engineering, is that we might only be interested in tracking the signal process $\{S(t)\}$ defined by (3), rather than the full state process $\{\mathbf{X}(t)\}$. Even so, if we want to arrive at a recursive filter, we should postulate a model for generating $\{S(t)\}$ of the form (1)–(3), that is, we should represent $\{S(t)\}$ as a projection of a vector Markov process. Further, our filter will still have to track the full state $\mathbf{X}(t)$, even though the estimate of $S(t)$ at each time t is all we want. Clearly, taking conditional expectation of both sides of (3), we have

$$\hat{S}(t) = E\{S(t)|\boldsymbol{\zeta}(t)\} = E\{\mathbf{h}^T(t)\mathbf{X}(t)|\boldsymbol{\zeta}(t)\}$$

$$= \mathbf{h}^T(t)E\{\mathbf{X}(t)|\boldsymbol{\zeta}(t)\} = \mathbf{h}^T(t)\hat{\mathbf{X}}(t) \tag{33}$$

Thus, if we know $\hat{\mathbf{X}}(t)$ we can easily find $\hat{S}(t)$.

Returning now to the issue of finding the propagate formula, we merely take the conditional expectation of both sides of (1):

$$E\{\mathbf{X}(t+1)|\boldsymbol{\zeta}(t)\} = E\{\mathbf{A}(t)\mathbf{X}(t) + \mathbf{B}(t)\mathbf{V}(t)|\boldsymbol{\zeta}(t)\}$$

$$= \mathbf{A}(t)E\{\mathbf{X}(t)|\boldsymbol{\zeta}(t)\} + \mathbf{B}(t)E\{\mathbf{V}(t)|\boldsymbol{\zeta}(t)\} \quad (34)$$

Now by definition, $\{\mathbf{V}(t)\}$ is a WGN sequence, independent of $\mathbf{X}(0)$. From equation (34) of Chapter 10, $\mathbf{X}(t)$ depends on $\mathbf{V}(0), \mathbf{V}(1), \ldots, \mathbf{V}(t-1)$, but not $\mathbf{V}(t)$. $Y(t)$ depends on $\mathbf{X}(t)$ and $N(t)$, but by assumption $N(t)$ is independent of $\mathbf{V}(t)$. Thus, $Y(1), Y(2), \ldots, Y(t)$ are all independent of $\mathbf{V}(t)$, that is, the data vector $\mathbf{Z}(t)$ is independent of $\mathbf{V}(t)$. As we have just emphasized, $\boldsymbol{\zeta}(t)$ is obtained *causally* from $\mathbf{Z}(t)$, so $\boldsymbol{\zeta}(t)$ is also independent of the particular noise input $\mathbf{V}(t)$. Hence we have

$$E\{\mathbf{V}(t)|\boldsymbol{\zeta}(t)\} = E\{\mathbf{V}(t)\} = 0 \quad (35)$$

Consequently, (34) reduces to

$$E\{\mathbf{X}(t+1)|\boldsymbol{\zeta}(t)\} = \mathbf{A}(t)E\{\mathbf{X}(t)|\boldsymbol{\zeta}(t)\} \quad (36)$$

which is the desired propagate formula. Since this derivation is valid for every sampling instant, we may reduce the time argument by one unit throughout (36) and still have a valid equation:

$$E\{\mathbf{X}(t)|\boldsymbol{\zeta}(t-1)\} = \mathbf{A}(t-1)E\{\mathbf{X}(t-1)|\boldsymbol{\zeta}(t-1)\} \quad (37)$$

The Kalman Filter Equations

Now write $\hat{\mathbf{X}}(t) = E\{\mathbf{X}(t)|\boldsymbol{\zeta}(t)\}$ and $\hat{\mathbf{X}}(t-1) = E\{\mathbf{X}(t-1)|\boldsymbol{\zeta}(t-1)\}$ for brevity, and substitute (37) into (23) to obtain

$$\hat{\mathbf{X}}(t) = \mathbf{A}(t-1)\hat{\mathbf{X}}(t-1) + \boldsymbol{\alpha}_{tt}\frac{\nu(t)}{d_t} \quad (38)$$

Equation (38) is our fundamental Kalman filtering formula. It includes both of the operations of propagating the state forward from one sampling time to the next and updating the estimate to incorporate the latest data. As we see, it is truly recursive in the sense that the dimension of the state of the filter remains constant.

In order to complete the derivation, we need to refer to our model equations along with results of Appendix 1 to find $\boldsymbol{\alpha}_t$ and d_t in terms of model parameters. We also need to obtain some equations that provide a theoretical measure of the estimation error associated with $\mathbf{X}(t)$.

We proceed to the task by adapting the notation of Appendix 1 to the present context. It will be most convenient if we replace the index k in Appendix 1 by the time argument $t-1$. Thus, equation (A1.3) now becomes

$$\mathbf{C}_t = \begin{bmatrix} \mathbf{C}_{t-1} & \mathbf{v}_{t-1} \\ \mathbf{v}_{t-1}^T & r_{t-1} \end{bmatrix} \tag{39}$$

Recalling (8) and (6), we see that

$$\mathbf{v}_{t-1} = E\{\mathbf{Z}(t-1)Y(t)\} \tag{40}$$

and also that

$$r_{t-1} = E\{Y^2(t)\} \tag{41}$$

In the present notation, (A1.4) and (A1.5) now become, respectively,

$$\mathbf{L}_t = \begin{bmatrix} \mathbf{L}_{t-1} & \mathbf{0} \\ \mathbf{w}_{t-1} & 1 \end{bmatrix} \qquad \mathbf{D}_t = \begin{bmatrix} \mathbf{D}_{t-1} & \mathbf{0} \\ \mathbf{0}^T & d_t \end{bmatrix} \tag{42}$$

$$\begin{aligned} \mathbf{w}_{t-1} &= \mathbf{D}_{t-1}^{-1}\mathbf{L}_{t-1}^{-1}\mathbf{v}_{t-1} \\ d_t &= r_{t-1} - \mathbf{w}_{t-1}^T\mathbf{D}_{t-1}\mathbf{w}_{t-1} \end{aligned} \tag{43}$$

Lastly, (A1.6), which is a very important formula, now reads

$$\mathbf{L}_t^{-1} = \begin{bmatrix} \mathbf{L}_{t-1}^{-1} & 0 \\ -\mathbf{w}_{t-1}^T\mathbf{L}_{t-1}^T & 1 \end{bmatrix} \tag{44}$$

We can now explain why the $\{\nu(t)\}$ process is called the *innovation process*. Recall its definition in equations (10)–(12). From those equations we see that we can write

$$\begin{bmatrix} \boldsymbol{\zeta}(t-1) \\ \nu(t) \end{bmatrix} = \begin{bmatrix} \mathbf{L}_{t-1}^{-1} & 0 \\ -\mathbf{w}_{t-1}^T\mathbf{L}_{t-1}^{-1} & 1 \end{bmatrix} \begin{bmatrix} \mathbf{Z}(t-1) \\ Y(t) \end{bmatrix} \tag{45}$$

The bottom equation here states that

$$\nu(t) = Y(t) - \mathbf{w}_{t-1}^T\mathbf{L}_{t-1}^{-1}\mathbf{Z}(t-1) \tag{46}$$

From (43), this may be written

$$\nu(t) = Y(t) - \mathbf{v}_{t-1}^T (\mathbf{L}_{t-1}^{-1})^T \mathbf{D}_{t-1}^T \mathbf{L}_{t-1}^{-1} \mathbf{Z}(t-1) \tag{47}$$

Making use of the fact that the transpose of the inverse is the same as the inverse of the transpose, along with (9), (47) becomes

$$\nu(t) = Y(t) - \mathbf{v}_{t-1}^T \mathbf{C}_{t-1}^{-1} \mathbf{Z}(t-1) \tag{48}$$

Making use of (16) and (40), we see that

$$\mathbf{v}_{t-1}^T \mathbf{C}_{t-1}^{-1} \mathbf{Z}(t-1) = E\{Y(t)|\mathbf{Z}(t-1)\} \tag{49}$$

Therefore, the formula (45), which follows from the algorithm in Appendix 1 for triangular factorization of a covariance matrix and which we have used to define the innovations, is equivalent to the condition

$$\nu(t) = Y(t) - E\{Y(t)|\mathbf{Z}(t-1)\} \tag{50}$$

From our discussion of conditional expectation, we know that $E\{Y(t)|\mathbf{Z}(t-1)\}$ is the optimum prediction of $Y(t)$ given the previous data $Y(1), Y(2), \ldots, Y(t-1)$. Therefore $\nu(t)$ is the difference between the actual value of $Y(t)$ and the best prediction of it given the previous data. It therefore represents that part of $Y(t)$ which is really *new* information. This is why it is customary to use the name *innovation* and the symbol ν.

Now refer to equations (4) and (33). Since $N(t)$ is independent of $\mathbf{Z}(t-1)$, then

$$E\{N(t)|\mathbf{Z}(t-1)\} = 0 \tag{51}$$

Therefore

$$E\{Y(t)|\mathbf{Z}(t-1)\} = \mathbf{h}^T(t) E\{\mathbf{X}(t)|\mathbf{Z}(t-1)\} \tag{52}$$

Recall that we are using exactly the same information whether we condition on $\mathbf{Z}(t-1)$ or $\boldsymbol{\zeta}(t-1)$; it is simply a question of the form taken by the equations. Either applying this statement to (37) directly, or repeating the argument that led to (37) using $\mathbf{Z}(t-1)$ rather than $\boldsymbol{\zeta}(t-1)$, as you prefer, one has

$$E\{\mathbf{X}(t)|\mathbf{Z}(t-1)\} = \mathbf{A}(t-1) E\{\mathbf{X}(t-1)|\mathbf{Z}(t-1)\} \tag{53}$$

From (50), (52), (53), and (7) we now obtain

$$\nu(t) = Y(t) - \mathbf{h}^T(t)\mathbf{A}(t-1)\hat{\mathbf{X}}(t-1) \tag{54}$$

Therefore, our Kalman filter formula may be rewritten

$$\hat{\mathbf{X}}(t) = \mathbf{A}(t-1)\hat{\mathbf{X}}(t-1) + \frac{1}{d_t}\boldsymbol{\alpha}_{tt}\left[Y(t) - \mathbf{h}^T(t)\mathbf{A}(t-1)\hat{\mathbf{X}}(t-1)\right] \quad (55)$$

In this form, the filter is now self-contained and expressed in a completely recursive way: The new estimate $\hat{\mathbf{X}}(t)$ involves only the previous estimate $\hat{\mathbf{X}}(t-1)$ (which summarizes all of the past data) combined with the current data point $Y(t)$.

It still remains to determine the two quantities $\boldsymbol{\alpha}_{tt}$ and d_t in terms of system model parameters. From (13), we have

$$d_t = E\{\nu^2(t)\} \quad (56)$$

From (20), we have

$$\boldsymbol{\alpha}_{tt} = E\{\mathbf{X}(t)\nu(t)\} \quad (57)$$

Our task now is to utilize (54) in (56) and (57) to come out with a usable way of calculating the *filter gain vector* $\mathbf{k}(t)$, which we define as

$$\mathbf{k}(t) = \frac{1}{d_t}\boldsymbol{\alpha}_{tt} \quad (58)$$

Toward this end it is useful to define the one-step prediction error $\boldsymbol{\Delta}(t)$ by

$$\boldsymbol{\Delta}(t) = \mathbf{X}(t) - E\{\mathbf{X}(t)|\mathbf{Z}(t-1)\} \quad (59)$$

We also define the current estimation error $\boldsymbol{\theta}(t)$ by

$$\boldsymbol{\theta}(t) = \mathbf{X}(t) - E\{\mathbf{X}(t)|\mathbf{Z}(t)\} \quad (60)$$

It is evident that the means of both of these are identically zero. Let us define the associated covariance matrices

$$\mathbf{M}(t) = E\{\boldsymbol{\Delta}(t)\boldsymbol{\Delta}^T(t)\} \quad (61)$$

$$\mathbf{P}(t) = E\{\boldsymbol{\theta}(t)\theta^T(t)\} \quad (62)$$

One of the exercises at the end of this chapter is to prove the following identities:

$$\begin{aligned} E\{\boldsymbol{\Delta}(t)E\{\mathbf{X}^T(t)|\mathbf{Z}(t-1)\}\} &= 0 \\ E\{\boldsymbol{\theta}(t)E\{\mathbf{X}^T(t)|\mathbf{Z}(t)\}\} &= 0 \end{aligned} \quad (63)$$

From (61)–(63) it follows that

$$\begin{aligned} \mathbf{M}(t) &= E\{\boldsymbol{\Delta}(t)\mathbf{X}^T(t)\} = E\{\mathbf{X}(t)\boldsymbol{\Delta}^T(t)\} \\ \mathbf{P}(t) &= E\{\boldsymbol{\theta}(t)\mathbf{X}^T(t)\} = E\{\mathbf{X}(t)\boldsymbol{\theta}^T(t)\} \end{aligned} \tag{64}$$

The second equation on each line is a consequence of the known facts that $\mathbf{M}(t)$ and $\mathbf{P}(t)$ are symmetric.

Now from (50) and (52) we have

$$\nu(t) = Y(t) - \mathbf{h}^T(t)E\{\mathbf{X}(t)|\mathbf{Z}(t-1)\} \tag{65}$$

Making use of (3), (4), and (59) one finds

$$\begin{aligned} \nu(t) &= \mathbf{h}^T(t)\boldsymbol{\Delta}(t) + N(t) \\ &= \boldsymbol{\Delta}^T(t)\mathbf{h}(t) + N(t) \end{aligned} \tag{66}$$

Substitute (66) into (57), recall that $\mathbf{X}(t)$ and $N(t)$ are independent, and use (64) to obtain

$$\boldsymbol{\alpha}_{tt} = \mathbf{M}(t)\mathbf{h}(t) \tag{67}$$

Observe that $\boldsymbol{\Delta}(t)$ and $N(t)$ are independent and use (66) in (56) along with (5) and (61) to obtain

$$d_t = \mathbf{h}^T(t)\mathbf{M}(t)\mathbf{h}(t) + R(t) \tag{68}$$

Therefore, if we have a way of computing the covariance matrix $\mathbf{M}(t)$, we can find the filter gain vector $\mathbf{k}(t)$ defined in (58).

Actually, all of the required relationships are contained in the algorithm from Appendix 1. However, at this point the following procedure is perhaps the most expedient. Make use of (1) and (53) to rewrite (59):

$$\begin{aligned} \boldsymbol{\Delta}(t) &= \mathbf{A}(t-1)\mathbf{X}(t-1) + \mathbf{B}(t-1)\mathbf{V}(t-1) \\ &\quad -\mathbf{A}(t-1)E\{\mathbf{X}(t-1)|\mathbf{Z}(t-1)\} \\ &= \mathbf{A}(t-1)\boldsymbol{\theta}(t-1) + \mathbf{B}(t-1)\mathbf{V}(t-1) \end{aligned} \tag{69}$$

Now use (61), (62), and (2) to obtain

$$\mathbf{M}(t) = \mathbf{A}(t-1)\mathbf{P}(t-1)\mathbf{A}^T(t-1) + \mathbf{B}(t-1)\mathbf{Q}(t-1)\mathbf{B}^T(t-1) \tag{70}$$

Our next task is to obtain an equation connecting $\mathbf{P}(t)$ and $\mathbf{M}(t)$, which will then provide an iterative method of computing $\mathbf{M}(t)$, as we will explain. To obtain our desired equation, subtract both sides of (23) from $\mathbf{X}(t)$:

$$\mathbf{X}(t) - E\{\mathbf{X}(t)|\zeta(t)\} = \mathbf{X}(t) - E\{\mathbf{X}(t)|\zeta(t-1)\} - \boldsymbol{\alpha}_{tt}\frac{\nu(t)}{d_t} \quad (71)$$

Recalling that conditioning on $\zeta(t)$ is equivalent statistically to conditioning on $\mathbf{Z}(t)$ permits us to write (71) as

$$\boldsymbol{\theta}(t) = \boldsymbol{\Delta}(t) - \boldsymbol{\alpha}_{tt}\frac{\nu_t}{d_t} \quad (72)$$

Multiply (72) on the right by $\mathbf{X}^T(t)$ and take the expected value. From (57) and (64), we get

$$\mathbf{P}(t) = \mathbf{M}(t) - \frac{1}{d_t}\boldsymbol{\alpha}_{tt}\boldsymbol{\alpha}_{tt}^T \quad (73)$$

Substituting (67) and (68) into (73) yields

$$\mathbf{P}(t) = \mathbf{M}(t) - [\mathbf{h}^T(t)\mathbf{M}(t)\mathbf{h}(t) + R(t)]^{-1}\mathbf{M}(t)\mathbf{h}(t)\mathbf{h}^T(t)\mathbf{M}(t) \quad (74)$$

Using the Filter

We have now completed the derivation of all of the Kalman filtering equations. Let us now explain briefly how a user actually implements the filter. The first step is to calculate the sequence of covariance matrices, $\mathbf{M}(1), \mathbf{M}(2), \ldots, \mathbf{M}(t), \ldots,$ for as many time points as there are for which one expects to receive data. This sequence can be precomputed off-line, before any data is actually received. The initial state of the plant was assumed to be random with mean $\mathbf{0}$ and covariance $\mathbf{P}_0$. By equation (70) we have

$$\mathbf{M}(1) = \mathbf{A}(0)\mathbf{P}_0\mathbf{A}^T(0) + \mathbf{B}(0)\mathbf{Q}(0)B^T(0) \quad (75)$$

This is the covariance of the error associated with making a one-step prediction of $\mathbf{X}(1)$ based on only the a priori information about $\mathbf{X}(0)$. The first data point that will be received is $Y(1)$. The theoretical error covariance associated with the estimate obtained by updating the one-step prediction by incorporating $Y(1)$ is, by definition, $\mathbf{P}(1)$. The reduction in error caused by incorporating the new data is seen by utilizing (74), which particularizes to

$$\mathbf{P}(1) = \mathbf{M}(1) - [\mathbf{h}^T(1)\mathbf{M}(1)\mathbf{h}(1) + R(1)]^{-1}\mathbf{M}(1)\mathbf{h}(1)\mathbf{h}^T(1)\mathbf{M}(1)$$

The user now proceeds to use (70) and (74) alternately, to compute iteratively the sequence of matrices $\{\mathbf{M}(t)\}$ as far as required. Once that is done, the associated sequence of Kalman filter gains $\{\mathbf{k}(t)\}$ is found, from (58), (67), and (68):

$$\mathbf{k}(t) = \mathbf{M}(t)\mathbf{h}(t)[\mathbf{h}^T\mathbf{M}(t)\mathbf{h}(t) + R(t)]^{-1} \tag{76}$$

At this point the precomputation is completed and the filter is ready for operating in real time as the data arrives. Note that, because the data sequence $\{Y(t)\}$ is scalar-valued, there is no matrix inversion, but only division by a scalar.

The first estimate is seen by definition to be

$$\mathbf{X}(0) = \mathbf{0} \tag{77}$$

since it is based only on a priori information. From there on, (55) is used to propagate and update simultaneously the estimate as each new data point is acquired:

$$\hat{\mathbf{X}}(t) = \mathbf{A}(t-1)\hat{\mathbf{X}}(t-1) + \mathbf{k}(t)[Y(t) - \mathbf{h}^T(t)\mathbf{A}(t-1)\hat{\mathbf{X}}(t-1)] \tag{78}$$

The sequence of matrices $\{\mathbf{P}(t)\}$ which was found as a by-product of the procedure for calculating $\{\mathbf{k}(t)\}$ is seen, from the definitions (60) and (62), to provide a theoretical measure of the error associated with each estimate.

As an exercise to test his or her understanding, the reader should determine how the results of this chapter need to be modified if the initial state has a nonzero mean **M**.

This completes our discussion of discrete-time Kalman filtering. We have derived the equations in only one of a number of possible equivalent forms. The reader will find alternate forms in Anderson and Moore (1979), Balakrishnan (1984), and the references listed therein. Also, we have done the derivation only for the case of scalar observations. The required extension for vector-valued observations is not difficult and will also be found in these references. Finally, the whole issue of finding the most numerically suitable procedure for computer implementation has not even been discussed. As stated at the beginning of this chapter, discussion of all of these issues really requires a separate course. For an introduction to these considerations, see Bierman (1977) and Orfandis (1985).

The main goal here has been to emphasize the intimate connections among the algorithm in Appendix 1, the Kalman filter itself, the innovations process, and the concepts of causality, invertibility, and recursivity. We trust this goal has been achieved to the readers' satisfaction.

Problems

1. Given the matrix

$$\mathbf{C}_4 = \begin{bmatrix} 1 & 1 & 1 & 1 \\ 1 & 2 & 4 & 8 \\ 1 & 4 & 16 & 64 \\ 1 & 8 & 64 & 512 \end{bmatrix}$$

let $\mathbf{C}_k$, for $k = 1, 2, 3$, denote the upper left-hand corner $k \times k$ submatrix of $\mathbf{C}_4$. Use the algorithm of Appendix 1 to factor each submatrix as $\mathbf{C}_k = \mathbf{L}_k \mathbf{D}_k \mathbf{L}_k^T$, and obtain the sequence of inverse matrices $\mathbf{C}_1^{-1}, \mathbf{C}_2^{-1}, \mathbf{C}_3^{-1}, \mathbf{C}_4^{-1}$.

2. Given a Gaussian random process $\{Y(t),\ t = 0, 1, 2, \ldots\}$, with $E\{Y(t)\} = 0$ for each t and $E\{Y(t)Y(s)\} = 2^{-|t-s|}$, for $t, s \geq 0$, compute

$$E\{Y(2)|Y(1)\},\ E\{Y(3)|Y(2), Y(1)\},\ E\{Y(4)|Y(3), Y(2), Y(1)\},$$

$$\text{and } E\{Y(5)|Y(4), Y(3), Y(2), Y(1)\}$$

3. Let $\mathbf{X}$ and $\mathbf{Z}$ be random vectors of dimension n and m, respectively. Let them have an arbitrary known joint distribution. Define

$$\boldsymbol{\Delta} = \mathbf{X} - E\{\mathbf{X}|\mathbf{Z}\}$$

Use the following two properties of conditional expectation

1. $$E\{\mathbf{X}\} = E\{E\{\mathbf{X}|\mathbf{Z}\}\}$$

2. $$E\{\mathbf{X}E\{\mathbf{X}^T|\mathbf{Z}\}|\mathbf{Z}\} = E\{\mathbf{X}|\mathbf{Z}\}E\{\mathbf{X}^T|\mathbf{Z}\}$$

to prove that

$$E\{\boldsymbol{\Delta}E\{\mathbf{X}^T|\mathbf{Z}\}\} = \mathbf{0}$$

This result is sometimes called the *Orthogonality Principle*. Show that this establishes the identities given in (63).

4. Given the discrete-time scalar signal model

$$X(t+1) = \tfrac{1}{2}X(t) + V(t), \qquad t = 0, 1, 2, \ldots$$

where $X(0) = 0$ and $\{V(\cdot)\}$ is unit zero mean WGN.

Given the measurement model

$$Y(t) = X(t) + N(t)$$

where $\{N(\cdot)\}$ is another zero mean unit WGN sequence, independent of $\{V(\cdot)\}$:

a. Write out the Kalman filter equations for this case.

b. By elimination and substitution, obtain an equation of the form

$$\hat{X}(t) = f(t)\hat{X}(t-1) + g(t)Y(t)$$

Also show that

$$P(t) = \frac{4 + P(t-1)}{8 + P(t-1)}$$

c. Show that

$$\lim_{t \to \infty} P(t) = \frac{\sqrt{65} - 7}{2} \approx 0.537$$

regardless of the starting value $P(0)$.

d. Find $\bar{f} = \lim_{t \to \infty} f(t)$ and $\bar{g} = \lim_{t \to \infty} g(t)$.

5. Given the discrete-time linear stochastic system with noisy observations

$$\mathbf{X}(t+1) = \mathbf{A}\mathbf{X}(t) + \mathbf{B}\mathbf{V}(t)$$

$$Y(t) = \mathbf{h}^T\mathbf{X}(t) + N(t)$$

The initial state is random with mean zero and covariance Λ. The sequence $\{\mathbf{V}(\cdot)\}$ is vector WGN with mean zero and $E\{\mathbf{V}(t)\mathbf{V}^T(t)\} = \mathbf{I}$ for every t.

The sequence $\{N(\cdot)\}$ is scalar WGN with mean zero and $E\{N^2(t)\} = \sigma^2$ for every t. $\mathbf{X}(0)$, $\{\mathbf{V}(\cdot)\}$, and $\{N(\cdot)\}$ are mutually independent. Given that

$$\mathbf{A} = \begin{bmatrix} 0 & a \\ a & 0 \end{bmatrix}, \quad \mathbf{B} = \begin{bmatrix} 0 & 1 \\ -1 & 0 \end{bmatrix}, \quad \mathbf{h} = \begin{bmatrix} 1 \\ 1 \end{bmatrix}$$

a. Write out the Kalman filter equations for this case.

b. By elimination and substitution, obtain a single equation that relates the error covariance $\mathbf{P}(t)$ to the error covariance $\mathbf{P}(t-1)$.

c. Assume that $\lim_{t \to \infty} \mathbf{P}(t) = \mathbf{P}^*$. Assume that $\mathbf{P}^*$ has the form $\mathbf{P}^* = \begin{bmatrix} p & q \\ q & p \end{bmatrix}$, where p and q are unknowns. Solve for p and q in terms of a and σ^2.

d. Calculate the corresponding limit filter gain $\mathbf{k}^* = \lim_{t \to \infty} \mathbf{k}(t)$. Assume that for large t it is acceptable to use the filter with constant gain $\mathbf{k}^*$ as an approximation to the exact Kalman filter. Write the equation for the constant gain filter if $a = \frac{1}{2}$ and $\sigma^2 = 4$.

Appendix 1

Triangular Factorization of Covariance Matrices

We begin from the well-known fact that any square matrix $\mathbf{A}$ can be written as a product of a lower triangular matrix $\mathbf{L}$ and an upper triangular matrix $\mathbf{U}$ in an infinite number of ways (Gerald, 1978):

$$\mathbf{A} = \mathbf{L}\mathbf{U} \tag{A1.1}$$

A common way of making the factorization unique, as noted by Gerald, is to require that $\mathbf{U}$ have only 1's on its main diagonal. However, if we are dealing with a positive definite symmetric matrix to begin with, we may proceed further. The following theorem presents a useful algorithm for that case.

Theorem. Let $\mathbf{C}$ be an $n \times n$ positive definite symmetric matrix. Then there exist (a) a unique matrix $\mathbf{L}_n$ which is lower triangular with all 1's on its main diagonal and (b) a unique matrix $\mathbf{D}_n$ which is diagonal such that

$$\mathbf{C} = \mathbf{L}_n\mathbf{D}_n\mathbf{L}_n^T \tag{A1.2}$$

Moreover, the matrices $\mathbf{L}_n$ and $\mathbf{D}_n$ can be found by the following sequential procedure.

PROCEDURE. For each integer k, $1 \leq k \leq n$, let $\mathbf{C}_k$ denote the upper left-hand $k \times k$ submatrix of $\mathbf{C}$. Define the vector $\mathbf{v}_k$ and the scalar r_k from the elements of the matrix $\mathbf{C}_{k+1}$:

$$\mathbf{C}_{k+1} = \begin{bmatrix} \mathbf{C}_k & \mathbf{v}_k \\ \mathbf{v}_k^T & r_k \end{bmatrix} \tag{A1.3}$$

Also let $\mathbf{L}_k$ denote a $k \times k$ lower triangular matrix with all 1's on its diagonal, and $\mathbf{D}_k$ a $k \times k$ diagonal matrix.

Define $\mathbf{L}_1 = 1$ and $\mathbf{D}_1 = c_{11}$. For $1 \le k \le n-1$, the matrices $\mathbf{L}_{k+1}$ and $\mathbf{D}_{k+1}$ are found from $\mathbf{L}_k$ and $\mathbf{D}_k$ by the following method:

$$\mathbf{L}_{k+1} = \begin{bmatrix} \mathbf{L}_k & \mathbf{0} \\ \mathbf{w}_k^T & 1 \end{bmatrix}, \qquad \mathbf{D}_{k+1} = \begin{bmatrix} \mathbf{D}_k & \mathbf{0} \\ \mathbf{0}^T & d_{k+1} \end{bmatrix} \tag{A1.4}$$

where

$$\begin{aligned} \mathbf{w}_k &= \mathbf{D}_k^{-1}\mathbf{L}_k^{-1}\mathbf{v}_k \\ d_{k+1} &= r_k - \mathbf{w}_k^T\mathbf{D}_k\mathbf{w}_k \end{aligned} \tag{A1.5}$$

Finally, in order to proceed to the next step, the inverse matrix $\mathbf{L}_{k+1}^{-1}$ is obtained by the rule:

$$\mathbf{L}_{k+1}^{-1} = \begin{bmatrix} \mathbf{L}_k^{-1} & \mathbf{0} \\ -\mathbf{w}_k^T\mathbf{L}_k^{-1} & 1 \end{bmatrix} \tag{A1.6}$$

Since the matrices $\mathbf{D}_k$ are diagonal, their inverses are readily found.

This completes the statement of the theorem. Before proceeding to its proof, we give an example of the algorithm.

Example

$$\mathbf{C} = \begin{bmatrix} 36 & 24 & 18 \\ 24 & 41 & 22 \\ 18 & 22 & 14 \end{bmatrix}$$

So then $\mathbf{C}_1 = c_{11} = 36$ and

$$\mathbf{C}_2 = \begin{bmatrix} 36 & 24 \\ 24 & 41 \end{bmatrix}$$

Hence $\mathbf{v}_1 = 24$ and $r_1 = 41$. By definition $\mathbf{L}_1 = 1$ and $\mathbf{D}_1 = 36$. Now

$$\mathbf{L}_2 = \begin{bmatrix} \mathbf{L}_1 & 0 \\ \mathbf{w}_1^T & 1 \end{bmatrix} = \begin{bmatrix} 1 & 0 \\ w_1 & 1 \end{bmatrix}$$

where

$$\mathbf{w}_1 = \mathbf{D}_1^{-1}\mathbf{L}_1^{-1}\mathbf{v}_1 = \tfrac{24}{36} = \tfrac{2}{3}$$

So

$$\mathbf{L}_2 = \begin{bmatrix} 1 & 0 \\ \frac{2}{3} & 1 \end{bmatrix}$$

Then

$$\begin{aligned} d_2 &= r_1 - \mathbf{w}_1^T \mathbf{D}_1 \mathbf{w}_1 \\ &= 41 - \left(\tfrac{2}{3}\right)^2 (36) \\ &= 25 \end{aligned}$$

Thus

$$\mathbf{D}_2 = \begin{bmatrix} 36 & 0 \\ 0 & 25 \end{bmatrix}$$

and by inspection,

$$\mathbf{D}_2^{-1} = \begin{bmatrix} \frac{1}{36} & 0 \\ 0 & \frac{1}{25} \end{bmatrix}$$

Also, $\mathbf{w}_1^T \mathbf{L}_1^{-1} = \frac{2}{3}$, so we have

$$\mathbf{L}_2^{-1} = \begin{bmatrix} 1 & 0 \\ -\frac{2}{3} & 1 \end{bmatrix}$$

Hence

$$\mathbf{w}_2 = \mathbf{D}_2^{-1}\mathbf{L}_2^{-1}\mathbf{v}_2 = \begin{bmatrix} \frac{1}{36} & 0 \\ 0 & \frac{1}{25} \end{bmatrix} \begin{bmatrix} 1 & 0 \\ -\frac{2}{3} & 1 \end{bmatrix} \begin{bmatrix} 18 \\ 22 \end{bmatrix} = \begin{bmatrix} \frac{1}{2} \\ \frac{2}{5} \end{bmatrix}$$

Next,

$$d_3 = r_2 - \mathbf{w}_2^T \mathbf{D}_2 \mathbf{w}_2$$

$$= 14 - \begin{bmatrix} \frac{1}{2} & \frac{2}{5} \end{bmatrix} \begin{bmatrix} 36 & 0 \\ 0 & 25 \end{bmatrix} \begin{bmatrix} \frac{1}{2} \\ \frac{2}{5} \end{bmatrix}$$

$$= 14 - \begin{bmatrix} \frac{1}{2} & \frac{2}{5} \end{bmatrix} \begin{bmatrix} 18 \\ 10 \end{bmatrix} = 14 - 9 - 4 = 1$$

Therefore

$$\mathbf{L}_3 = \begin{bmatrix} \mathbf{L}_2 & \mathbf{0} \\ \mathbf{w}_2^T & 1 \end{bmatrix} = \begin{bmatrix} 1 & 0 & 0 \\ \frac{2}{3} & 1 & 0 \\ \frac{1}{2} & \frac{2}{5} & 1 \end{bmatrix}$$

and

$$\mathbf{D}_3 = \begin{bmatrix} \mathbf{D}_2 & \mathbf{0} \\ \mathbf{0}^T & d_3 \end{bmatrix} = \begin{bmatrix} 36 & 0 & 0 \\ 0 & 25 & 0 \\ 0 & 0 & 1 \end{bmatrix}$$

and $\mathbf{C}_3 = \mathbf{L}_3 \mathbf{D}_3 \mathbf{L}_3^T$, as may be verified by direct calculation.

We now proceed to the proof of the theorem.

PROOF. The proof is carried out inductively on the index k. First of all, equations (A1.3) and (A1.4) are merely definitions, so they require no proof other than to observe that the indicated partitioning and labeling of the given matrices are always possible. We now show that for each value of k, the specified procedure does yield the result

$$\mathbf{C}_k = \mathbf{L}_k \mathbf{D}_k \mathbf{L}_k^T \tag{A1.7}$$

For $k = 1$, we have $\mathbf{L}_1 = 1$ and $\mathbf{D}_1 = c_{11}$ always, so (A1.7) clearly holds. Next,

$$\mathbf{C}_2 = \begin{bmatrix} c_{11} & v_1 \\ v_1 & r_1 \end{bmatrix}$$

by (A1.3), and by (A1.4) we have

$$\mathbf{L}_2 = \begin{bmatrix} 1 & 0 \\ w_1 & 1 \end{bmatrix}, \qquad \mathbf{D}_2 = \begin{bmatrix} c_{11} & 0 \\ 0 & d_2 \end{bmatrix}$$

Now perform the multiplication

$$\mathbf{L}_2\mathbf{D}_2\mathbf{L}_2^T = \begin{bmatrix} 1 & 0 \\ w_1 & 1 \end{bmatrix}\begin{bmatrix} c_{11} & 0 \\ 0 & d_2 \end{bmatrix}\begin{bmatrix} 1 & w_1 \\ 0 & 1 \end{bmatrix}$$

$$= \begin{bmatrix} 1 & 0 \\ w_1 & 1 \end{bmatrix}\begin{bmatrix} c_{11} & c_{11}w_1 \\ 0 & d_2 \end{bmatrix} = \begin{bmatrix} c_{11} & c_{11}w_1 \\ c_{11}w_1 & c_{11}w_1^2 + d_2 \end{bmatrix}$$

Since we wish to have $\mathbf{C}_2 = \mathbf{L}_2\mathbf{D}_2\mathbf{L}_2^T$, this requires that

$$v_1 = c_{11}w_1$$

$$r_1 = c_{11}w_1^2 + d_2$$

If c_{11}, v_1, and r_1 are given, we may compute w_1 and d_2 from the formulas

$$w_1 = \frac{v_1}{c_{11}}$$
$$d_2 = r_1 - c_{11}w_1^2 \tag{A1.8}$$

But since $\mathbf{L}_1 = 1$ and $\mathbf{D}_1 = c_{11}$, we see that (A1.8) is exactly what (A1.5) becomes for the case $k = 2$.

At this point it is necessary to recall a well-known result from matrix theory, namely, that the $n \times n$ matrix $\mathbf{C}$ is positive definite if and only if all of the so-called principal minors, that is, the determinants of the sub-matrices $|\mathbf{C}_k|$, are positive (Bellman, 1960, p. 116).

In particular, the hypothesis that $\mathbf{C}$ is positive definite insures that $c_{11} > 0$, so the first of equations (A1.8) is always meaningful. We now need also to show that it can never happen that $d_2 = 0$. To see this, substitute the first of equations (A1.8) into the second to obtain

$$d_2 = r_1 - c_{11}\left(\frac{v_1}{c_{11}}\right)^2 = r_1 - \frac{v_1^2}{c_{11}}$$

Now $d_2 = 0$ implies that

$$c_{11}r_1 = v_1^2$$

which would mean that the determinant of $\mathbf{C}_2$ is zero, contrary to hypothesis.

Therefore we have verified that if we make the definitions (A1.3) and (A1.4) for $k = 1$, and then apply (A1.5), we will indeed obtain matrices $\mathbf{L}_2$ and $\mathbf{D}_2$ such that $\mathbf{C}_2 = \mathbf{L}_2\mathbf{D}_2\mathbf{L}_2^T$.

Now suppose that for any given integer k we already have $\mathbf{C}_K = \mathbf{L}_k\mathbf{D}_k\mathbf{L}_k^T$. Consider the product

$$\mathbf{L}_{k+1}\mathbf{D}_{k+1}\mathbf{L}_{k+1}^T = \begin{bmatrix} \mathbf{L}_k & \mathbf{0} \\ \mathbf{w}_k^T & 1 \end{bmatrix} \begin{bmatrix} \mathbf{D}_k & \mathbf{0} \\ \mathbf{0}^T & d_{k+1} \end{bmatrix} \begin{bmatrix} \mathbf{L}_k^T & \mathbf{w}_k \\ \mathbf{0}^T & 1 \end{bmatrix}$$

$$= \begin{bmatrix} \mathbf{L}_k & \mathbf{0} \\ \mathbf{w}_k^T & 1 \end{bmatrix} \begin{bmatrix} \mathbf{D}_k\mathbf{L}_k^T & \mathbf{D}_k\mathbf{w}_k \\ \mathbf{0}^T & d_{k+1} \end{bmatrix}$$

$$= \begin{bmatrix} \mathbf{L}_k\mathbf{D}_k\mathbf{L}_k^T & \mathbf{L}_k\mathbf{D}_k\mathbf{w}_k \\ \mathbf{w}_k^T\mathbf{D}_k\mathbf{L}_k^T & \mathbf{w}_k^T\mathbf{D}_k\mathbf{w}_K + d_{k+1} \end{bmatrix}$$

$$= \begin{bmatrix} \mathbf{C}_k & \mathbf{v}_k \\ \mathbf{v}_k^T & r_k \end{bmatrix}$$

Equating the submatrices of the partitioned matrices and solving for $\mathbf{w}_k$ and d_{k+1} yields (A1.5). Therefore, if for some given integer k we already have $\mathbf{C}_k = \mathbf{L}_k\mathbf{D}_k\mathbf{L}_k^T$, and if we are given $\mathbf{C}_{k+1}$ and we make the definitions (A1.3) and (A1.4), then the application of (A1.5) will yield matrices $\mathbf{L}_{k+1}$ and $\mathbf{D}_{k+1}$ such that

$$\mathbf{C}_{k+1} = \mathbf{L}_{k+1}\mathbf{D}_{k+1}\mathbf{L}_{k+1}^T$$

Therefore, by induction, (A1.7) will hold for every value of k, $k = 1, 2, \ldots, n$. Thus, in particular, we arrive at

$$\mathbf{C} = \mathbf{L}_n\mathbf{D}_n\mathbf{L}_n^T$$

which is what was asserted in (A1.2). The uniqueness follows from the fact that the identification of submatrices given by (A1.5) is unique.

In order to complete the proof we need to show that the matrix $\mathbf{D}_k$ will never be singular for any k. In turn, that will ensure that Eq. (A1.5) is always meaningful. Incidentally, note that it is automatic that $\mathbf{L}_k^{-1}$ always exists, because $\mathbf{L}_k$ is lower triangular with all 1's on its diagonal, by construction, hence its determinant is always equal to unity.

We have already shown that $\mathbf{D}_1$ and $\mathbf{D}_2$ are always nonsingular. We proceed by induction. Assuming that $\mathbf{D}_k$ is nonsingular, we show that d_{k+1} can never be zero.

From (A1.5) we have

$$\mathbf{w}_k^T \mathbf{D}_k \mathbf{w}_k = \mathbf{v}_k^T (\mathbf{L}_k^{-1})^T \mathbf{D}_k^{-1} \mathbf{L}_k^{-1} \mathbf{v}_k$$

$$= \mathbf{v}_k^T \mathbf{C}_k^{-1} \mathbf{v}_k$$

The necessary and sufficient condition that $\mathbf{C}$ be positive definite, mentioned earlier, guarantees that $\mathbf{C}_k^{-1}$ exists.

In order for $d_{k+1} = 0$, it would have to happen that

$$r_k = \mathbf{w}_k^T \mathbf{D}_k \mathbf{w}_k = \mathbf{v}_k^T \mathbf{C}_k^{-1} \mathbf{v}_k$$

Now consider (A1.3). Define the k-dimensional vector $\mathbf{u}_k$ by

$$\mathbf{u}_k = -\mathbf{C}_k^{-1} \mathbf{v}_k$$

We then have

$$\mathbf{C}_{k+1} \begin{bmatrix} \mathbf{u}_k \\ 1 \end{bmatrix} = \begin{bmatrix} \mathbf{C}_k & \mathbf{v}_k \\ \mathbf{v}_k^T & r_k \end{bmatrix} \begin{bmatrix} \mathbf{u}_k \\ 1 \end{bmatrix}$$

$$= \begin{bmatrix} \mathbf{C}_k \mathbf{u}_k + \mathbf{v}_k \\ \mathbf{v}_k^T \mathbf{u}_k + r_k \end{bmatrix} = \begin{bmatrix} -\mathbf{v}_k + \mathbf{v}_k \\ -\mathbf{v}_k^T \mathbf{C}_k^{-1} \mathbf{v}_k + r_k \end{bmatrix}$$

Therefore, if it ever occurs that $r_k = \mathbf{v}_k^T \mathbf{C}_k^{-1} \mathbf{v}_k$, we could construct a $(k+1)$-dimensional vector, namely, $\begin{bmatrix} \mathbf{u}_k \\ 1 \end{bmatrix}$, such that

$$\mathbf{C}_{k+1} \begin{bmatrix} \mathbf{u}_k \\ 1 \end{bmatrix} = \begin{bmatrix} \mathbf{0} \\ 0 \end{bmatrix}$$

This would mean that a nonzero vector lies in the null space of $\mathbf{C}_{k+1}$, hence that $\mathbf{C}_{k+1}$ has rank less than $k+1$, which is impossible under the hypothesis that $\mathbf{C}$ is positive definite. Thus, the condition $d_{k+1} = 0$ would contradict this hypothesis, hence it can never happen that $d_{k+1} = 0$.

Appendix 2

Statistics of Monte Carlo Simulation

A Random Number Generator

We present here a FORTRAN program segment for a simple and crude random number generator. Its performance is installation-specific, because the performance depends upon how the particular operating system handles overflow. Also, there are certainly generators available which give superior results when subjected to standard statistical tests for Gaussianness, randomness, and so on. Nevertheless, the performance of the program given here is quite adequate to illustrate the principles discussed in Chapter 4.

The following program segment will generate 1000 pseudorandom numbers with an approximate Gaussian distribution having mean zero and variance one and store them in an array called GAUSS. It is intended that this segment will be embedded in a larger program to be created by the reader.

```
DIMENSION    GAUSS(1000)
TWOPI = 6.2831853
N = 16 807
T = 2.0* *(-31)
L = 5
DO 100 J = 1,1000
    K = N*L
    L = IABS(K)
    X = L
    RAND = X*T
    KTEST = (J + 1)/2 - J/2 + 1
    GO TO (90,80), KTEST
```

```
80          U1 = SQRT(ABS(-2.0*ALOG(RAND)))
            GO TO 100
90          U2 = TWOPI*RAND
            GAUSS (J - 1) = U1*COS(U2)
            GAUSS (J) = U1*SIN(U2)
100     CONTINUE
```

We will soon present some results obtained by using this program and discuss them with regard to some of the statements made in Chapter 4. In order to do that, however, it is necessary first to present a few results from the subject of Statistics.

Estimation of the Parameters of a Distribution

Suppose we are concerned with some scalar-valued r.v. X, having a probability density function $f_X(x;\alpha,\beta)$, where α and β are numerically valued parameters. Assume that we can make N repeated measurements of X, which we denote $X_1, X_2, \ldots, X_N$. We now present a classical technique for estimating the values of the parameters α and β, by making calculations using $X_1, X_2, \ldots, X_N$.

This technique is called the *Method of Maximum Likelihood*. Define the log-likelihood function

$$L(x_1, x_2, \ldots, x_N; \alpha, \beta) = \log\left(\prod_{i=1}^{N} f_X(x_i; \alpha, \beta)\right) = \sum_{i=1}^{N} \log f(x_i; \alpha, \beta) \tag{A2.1}$$

Here $x_1, x_2, \ldots, x_N$ are N independent variables used as arguments in the density function.

According to the method, the appropriate estimates are the roots of the so-called *likelihood equations*, which are obtained by computing $\partial L/\partial\alpha$ and $\partial L/\partial\beta$, equating them to zero, and then substituting $x_i = X_i$, $i = 1, \ldots, N$. The desired estimates are the values of α and β which satisfy these equations.

Specifically, let f_X be a Gaussian density with mean μ and variance σ^2. Let us find the maximum likelihood estimates of μ and σ^2. For this case, (A2.1) becomes

$$L(x_1, x_2, \ldots, x_N; \mu, \sigma^2) = -\frac{N}{2}\log\sigma^2 - \frac{N}{2}\log(2\pi) - \frac{1}{2\sigma^2}\sum_{i=1}^{N}(x_i - \mu)^2 \tag{A2.2}$$

Evaluating $\partial L/\partial\mu$, we find

$$\frac{\partial L}{\partial \mu} = \frac{1}{\sigma^2}\sum_{i=1}^{N} x_i - \frac{N}{\sigma^2}\mu \tag{A2.3}$$

Equating this to zero and solving for μ yields

$$\mu = \frac{1}{N}\sum_{i=1}^{N} x_i \tag{A2.4}$$

Substituting the actual observations X_i for the variables x_i provides our estimate:

$$\hat{\mu} = \frac{1}{N}\sum_{i=1}^{N} X_i \tag{A2.5}$$

Repeating the procedure for σ^2 produces the following results:

$$\frac{\partial L}{\partial \sigma^2} = -\frac{N}{2\sigma^2} + \frac{1}{2(\sigma^2)^2}\sum_{i=1}^{N}(x_i - \mu)^2 \tag{A2.6}$$

$$\frac{\partial L}{\partial \sigma^2} = 0 \Rightarrow \sigma^2 = \frac{1}{N}\sum_{i=1}^{N}(x_i - \mu)^2 \tag{A2.7}$$

To obtain the maximum likelihood estimator of σ^2, replace each x_i by X_i and μ by $\hat{\mu}$:

$$\hat{\sigma}^2 = \frac{1}{N}\sum_{i=1}^{N}(X_i - \hat{\mu})^2 \tag{A2.8}$$

According to Statistics, an important property for an estimator to possess is that of *unbiasedness*. If α is the true value of the parameter and $\hat{\alpha}$ is the estimator, then $\hat{\alpha}$ is called unbiased if

$$E\{\hat{\alpha}\} = \alpha \tag{A2.9}$$

Trying this for $\hat{\mu}$ yields

$$E\{\hat{\mu}\} = \frac{1}{N}E\left\{\sum_{i=1}^{N} X_i\right\} = \frac{1}{N}\sum_{i=1}^{N} E\{X_i\} = \frac{N\mu}{N} = \mu \tag{A2.10}$$

so $\hat{\mu}$ is unbiased.

Now consider:

$$(X_i - \hat{\mu})^2 = \left(X_i - \frac{1}{N} \sum_{j=1}^{N} X_j \right)^2 = X_i^2 - \frac{2X_i}{N} \sum_{j=1}^{N} X_j + \frac{1}{N^2} \left(\sum_{j=1}^{N} X_j \right)^2 \tag{A2.11}$$

$$\sum_{i=1}^{N} (X_i - \hat{\mu})^2 = \sum_{i=1}^{N} X_i^2 - \frac{2}{N} \left(\sum_{i=1}^{N} X_i \right) \left(\sum_{j=1}^{N} X_j \right) + \frac{N}{N^2} \left(\sum_{j=1}^{N} X_j \right)^2$$

$$= \sum_{i=1}^{N} X_i^2 - \frac{1}{N} \left(\sum_{i=1}^{N} X_i \right)^2 \tag{A2.12}$$

Since the X_i are mutually independent and each one is Gaussian with mean μ and variance σ^2, then if we call their sum Y,

$$Y = \sum_{i=1}^{N} X_i \tag{A2.13}$$

then Y is a Gaussian r.v. with mean $N\mu$ and variance $N\sigma^2$. Therefore

$$\begin{aligned} E\{X_i^2\} &= \mu^2 + \sigma^2, \qquad i = 1, 2, \ldots, N \\ E\{Y^2\} &= N^2\mu^2 + N\sigma^2 \end{aligned} \tag{A2.14}$$

We may use (A2.14) to take the expected value of (A2.12):

$$E\left\{ \sum_{i=1}^{N} (X_i - \mu)^2 \right\} = \sum_{i=1}^{N} E\{X_i^2\} - \frac{1}{N} E\{Y^2\}$$

$$= N\mu^2 + N\sigma^2 - N\mu^2 - \sigma^2 = (N-1)\sigma^2 \tag{A2.15}$$

Hence the maximum likelihood estimator $\hat{\sigma}^2$ given by (A2.8) is *not* unbiased:

$$E\{\hat{\sigma}^2\} = \frac{N-1}{N} \sigma^2 \neq \sigma^2 \tag{A2.16}$$

For this reason, it is common practice to replace (A2.8) by the so-called *sample variance*

$$V = \frac{1}{N-1} \sum_{i=1}^{N} (X_i - \hat{\mu})^2 \qquad \text{(A2.17)}$$

Now referring to the statements made in Chapter 4, the quantity designated M_2 in equation (9) of that chapter is just $\hat{\mu}$ for the case $N = 1000$ for the r.v. W_2. Similarly, V_2 in (10) of Chapter 4 is just V of (A2.17) above, for the same case. The principle of unbiasedness is the explanation for the divisor of 999 rather than 1000.

Statistics of Estimators

The next question that needs to be considered is, how large does the sample size N have to be in order to get a trustworthy estimate? In order to answer this question we will invoke the Central Limit Theorem (Ash, 1970). This theorem is concerned with the distribution of the scaled sum of a set of independent, identically distributed r.v.'s, as the number of terms in the sum becomes very large. It states that under appropriate regularity conditions, the distribution of the scaled sum will be Gaussian, regardless of the actual distribution of the underlying individual r.v.'s. Let us introduce the two r.v.'s

$$\begin{aligned} \Delta_1 &= \hat{\mu} - \mu \\ \Delta_2 &= V - \sigma^2 \end{aligned} \qquad \text{(A2.18)}$$

By the unbiasedness of $\hat{\mu}$ and V, Δ_1 and Δ_2 have zero means. Let us calculate their variances. From (A2.5) we have

$$\Delta_1 = \frac{1}{N} \sum_{i=1}^{N} (X_i - \mu) \qquad \text{(A2.19)}$$

Since the variance of a sum of independent r.v.'s is the sum of the variances, the variance of $N\Delta_1$ is $N\sigma^2$, that is,

$$E\{(N\Delta_1)^2\} = N^2 E\{\Delta_1^2\} = N\sigma^2 \qquad \text{(A2.20)}$$

whence

$$E\{\Delta_1^2\} = \frac{\sigma^2}{N} \qquad \text{(A2.21)}$$

The standard deviation of Δ_1 is thus $\sigma/\sqrt{N}$. Now, as is well known, the area under the Gaussian density between ± 3 standard deviations is .9974. Therefore, with a probability of .9974, the true value of μ is somewhere in the interval $[\hat{\mu} - 3\sigma/\sqrt{N}, \hat{\mu} + 3\sigma/\sqrt{N}]$.

The application of this result is as follows. Suppose the random number generator computer program given at the beginning of this appendix were used to generate 10,000 r.v.'s, rather than 1000 as it is written. Then $N = 10{,}000$, so $\sqrt{N} = 100$. If we calculate $\hat{\mu}$ in an attempt to verify that the parameter $\hat{\mu}$ for this generator really is zero, then all we can say is that with probability .9974, $\hat{\mu} - .03 \leq \mu \leq \hat{\mu} + .03$.

A similar calculation can be made for Δ_2, but it turns out to be rather tedious and the final result does not provide insight commensurate with the effort required to obtain it. In order to produce a figure for the number of samples required for the value of V to be a reliable estimate of σ^2, we return to (A2.7). If the true value of μ were actually known exactly, then the maximum likelihood estimate of σ^2 would be

$$V^{\#} = \frac{1}{N} \sum_{i=1}^{N} (X_i - \mu)^2 \tag{A2.22}$$

Furthermore, this statistic is unbiased with the divisor N as it stands. Let us replace Δ_2 in (A2.18) by

$$\Delta_2^{\#} = V^{\#} - \sigma^2 \tag{A2.23}$$

Again, $E\{\Delta_2^{\#}\} = 0$ and we seek $E\{(\Delta_2^{\#})^2\}$.

Since the terms of the sum in (A2.22) are mutually independent, in contrast to the terms in the sum in (A2.17), this is easy. We have

$$E\{(\Delta_2^{\#})^2\} = E\{(V^{\#})^2\} - \sigma^4 \tag{A2.24}$$

$$(V^{\#})^2 = \frac{1}{N^2} \sum_{i=1}^{N} (X_i - \mu)^4 + \frac{1}{N^2} \sum_{i=1}^{N} \sum_{\substack{j=1 \\ j \neq i}}^{N} (X_i - \mu)^2 (X_j - \mu)^2 \tag{A2.25}$$

By differentiating the characteristic function to evaluate the fourth central moment of a Gaussian distribution, one finds

$$E\{(X_i - \mu)^4\} = 3\sigma^4 \tag{A2.26}$$

In the double summation, when $i \neq j$ the factors are independent, so

$$E\{(X_i - \mu)^2 (X_j - \mu)^2\} = E\{(X_i - \mu)^2\} E\{(X_i - \mu)^2\} = \sigma^4 \quad \text{(A2.27)}$$

There is no problem interchanging the order of expectation and summation since these are only finite sums. Therefore

$$\begin{aligned} E\{(V^{\#})^2\} &= \frac{1}{N^2} \sum_{i=1}^{N} E\{(X_i - \mu)^4\} + \frac{1}{N^2} \sum_{i=1}^{N} \sum_{\substack{j=1 \\ j \neq i}}^{N} E\{(X_i - \mu)^2 (X_j - \mu)^2\} \\ &= \frac{N(3\sigma^4)}{N^2} + \frac{N(N-1)\sigma^4}{N^2} \\ &= \left(1 + \frac{2}{N}\right)\sigma^4 \end{aligned} \quad \text{(A2.28)}$$

We used the facts that there are N identical terms in the single summation and $N(N-1)$ identical terms in the double summation. The form of the final result in (A2.28) can be checked by observing that, for $N \to \infty$, $V^{\#}$ in (A2.22) converges to a sure variable of value σ^2, so its square will equal σ^4. Finally turning back to $\Delta^{\#}$, substituting (A2.28) into (A2.24) yields

$$E\{(\Delta^{\#})^2\} = \frac{2}{N}\sigma^4 \quad \text{(A2.29)}$$

Suppose we want to choose N in a Monte Carlo simulation so that the estimated value of σ^2 is within 1% of the true value, with probability .9974. The estimator we actually would use likely would be V. Since $V^{\#}$ is a more reliable estimator than V, the corresponding number N for $V^{\#}$ can be taken as a lower bound on the value of N required for V.

Our requirement is that 3 standard deviations of $\Delta^{\#}$ should equal $0.01\sigma^2$ or less. From (A2.29)

$$3\sqrt{E\{(\Delta^{\#})^2\}} = 3\sigma^2\sqrt{\frac{2}{N}} \leq 0.01\sigma^2 \quad \text{(A2.30)}$$

Rearranging and solving for N,

$$N \geq 18{,}000 \quad \text{(A2.31)}$$

Therefore, to simulate a Gaussian r.v. with mean zero and variance one, in order to check on the correctness of the simulation, we need 18,000 samples to be able to say with confidence that $V^{\#}$ should fall in the interval [.99, 1.01]. That is, if $V^{\#}$ does *not* fall in that interval we would definitely be justified in questioning the performance of the random number generator.

Finally, it needs to be stressed that these results apply to a single r.v. If one is concerned with simulation of dynamical systems, then these results apply to *each time point* on a trajectory. To achieve the same level of confidence regarding the validity of the simulation, it is necessary to generate 18,000 trajectories. Results obtained on the basis of a considerably smaller value of N may provide some psychological amelioration or be indicative of a trend, but one is not entitled to claim that the statistical theory of Monte Carlo simulation has thereby caused the abatement of all doubt.

Numerical Results of a Computer Experiment

The random number generator given at the beginning of this appendix was used to conduct a version of the computer experiment discussed in Chapter 4, involving discrete-time finite-length white Gaussian noise. In order to be able to create an ensemble of samples larger than 1000, we took the length (in time) of the white noise to be 6 rather than 100. Specifically, the computer generated many realizations of the vector

$$\mathbf{W}^T(\omega) = [W_1(\omega) \quad W_2(\omega) \quad W_3(\omega) \quad W_4(\omega) \quad W_5(\omega) \quad W_6(\omega)] \tag{A2.32}$$

The statistics **M** and **V**, as illustrated by equations (9) and (10) of Chapter 4, and in fact all of the elements of the covariance matrix, as illustrated by (11) of Chapter 4, were actually calculated for two cases: $N = 1000$ and $N = 10{,}000$.

The results are presented in Tables A2.1 and A2.2. The nominal values of μ and σ^2 were $\mu = 0$, $\sigma^2 = 1$. The statistic designated M in Chapter 4 was denoted $\hat{\mu}$ in (A2.5) of this appendix. We found the standard deviation of this around its true value is $\sigma/\sqrt{N}$. Taking $\sigma = 1$ and $N = 1000$ gives a standard deviation of 0.0316, and so 3 standard deviations correspond to ± 0.095 to achieve the .9974 confidence level. The variable farthest from the nominal value of zero is $M_3 = 0.05896$, which does fall within these limits.

We actually calculated the C_{ij} as indicated by equation (11) of Chapter 4, so the results correspond to V, given by equation (A2.17) of this appendix. The theoretical result (A2.29) found for $V^{\#}$ tends to underesti-

TABLE A2.1. Values of M_i and C_{ij} for $N = 1000$

M_1	M_2	M_3	M_4	M_5	M_6
−0.01288	0.01327	0.05896	−0.02172	0.00763	0.00523

C_{ij} (i ↓, j →)	1	2	3	4	5	6
1	1.04044	0.02652	−0.00438	−0.06036	−0.03187	−0.00409
2	0.02652	1.00571	0.01508	−0.01914	0.01080	−0.00452
3	−0.00438	0.01508	1.10369	0.05000	−0.03849	−0.05254
4	−0.06036	−0.01914	0.05000	1.00404	−0.07710	0.02317
5	−0.03187	0.01080	−0.03849	−0.07710	0.96641	−0.00890
6	−0.00409	−0.00452	0.05254	0.02317	0.00890	1.11825

mate N, as we discussed. Nevertheless, using $\sigma^2 = 1$ and $N = 1000$ in (A2.29) gives a standard deviation of 0.04472, and hence ± 3 standard deviations would be ± 0.1342. As we see in Table A2.1, $C_{66} = 1.11825$ lies the farthest from the nominal value of unity, but it falls within the above limit.

Repeating this discussion for the case $N = 10{,}000$ gives a standard deviation of 0.01 for M, so ± 3 standard deviations is ± 0.03. The largest value, $M_1 = 0.02071$, still falls within this bound.

For C_{ij}, 1 standard deviation is 0.01414, so ± 3 standard deviations is ± 0.04242. The sample of V with the largest deviation, $C_{11} = 1.03475$, still meets this limit.

TABLE A2.2. Values of M_i and C_{ij} for $N = 10{,}000$

M_1	M_2	M_3	M_4	M_5	M_6
0.02071	−0.01462	0.01525	−0.00581	−0.00524	0.00796

C_{ij} (i ↓, j →)	1	2	3	4	5	6
1	1.03475	0.00665	0.00581	−0.01892	−0.00740	0.00350
2	0.00665	0.98715	−0.00058	0.00507	0.00250	−0.00034
3	0.00581	−0.00058	1.01101	−0.01987	0.00643	0.01285
4	−0.01892	0.00507	−0.01987	0.99480	0.01149	−0.00889
5	−0.00740	0.00250	0.00643	0.01149	1.00543	−0.00329
6	0.00350	−0.00034	0.01285	−0.00889	−0.00329	1.01954

Even though the computer program for a random number generator given here is quite simple, on the basis of the above results alone we are obliged to conclude that it is performing satisfactorily.

At least in this appendix, if not always throughout the text, we have shown how the theory of probability and statistics combines with practice in an aleatory environment to provide the insight that is required for confident decision-making.

References

Anderson, B. and Moore, J. (1979). *Optimal Filtering*. Prentice-Hall, Englewood Cliffs, N.J.

Ash, R. B. (1970). *Basic Probability Theory*. John Wiley & Sons, New York.

Balakrishnan, A. V. (1981). *Applied Functional Analysis*, 2nd ed. Springer-Verlag, New York.

Balakrishnan, A. V. (1983). *Elements of State Space Theory of Systems*. Optimization Software, Inc., Publications Division (distributed by Springer-Verlag), New York.

Balakrishnan, A. V. (1984). *Kalman Filtering Theory*. Optimization Software, Inc., Publications Division (distributed by Springer-Verlag), New York.

Barrett, J. F. (1963). "The Use of Functionals in the Analysis of Nonlinear Physical Systems," *J. Electron. Control*, **15**(6), 567–615.

Barrett, J. F. (1964). "Hermite Functional Expansions and the Calculation of Output Autocorrelation and Spectrum for Any Time-Invariant Non-linear System with Noise Input," *J. Electron. Control*, **16**(1), 107–113.

Bellman, R. (1960). *Introduction to Matrix Analysis*. McGraw-Hill, New York.

Bierman, G. J. (1977). *Factorization Methods for Discrete Sequential Estimation*. Academic Press, New York.

Bremmerman, H. (1965). *Distributions, Complex Variables, and Fourier Transforms*. Addison-Wesley, Reading, MA.

Burdic, W. S. (1968). *Radar Signal Analysis*. Prentice-Hall, Englewood Cliffs, N.J.

Cook, C. E. and Bernfeld, M. (1967). *Radar Signals*. Academic Press, New York.

Courant, R. and Hilbert, D. (1953). *Methods of Mathematical Physics*, Vol. 1. Wiley-Interscience, New York.

Davenport, W. B., Jr. and Root, W. L. (1958). *Random Signals and Noise*. McGraw-Hill, New York.

Deutsch, R. (1962). *Nonlinear Transformations of Random Processes*. Prentice-Hall, Englewood Cliffs, N.J.

Franks, L. E. (1981). *Signal Theory*, revised ed. Dowden and Culver, Stroudsburg, PA.

Gerald, C. F. (1978). *Applied Numerical Analysis*, 2nd ed., p. 89. Addison-Wesley, Reading, MA.

Kailath, T. (1980). *Linear Systems*. Prentice-Hall, Englewood Cliffs, N.J.

Karlin, S. and Taylor, H. M. (1975). *A First Course in Stochastic Processes*, 2nd ed. Academic Press, New York.

Landau, H. and Pollack, H. (1961). "Prolate Spheroidal Wave Functions, Fourier Analysis, and Uncertainty—II," *Bell System Tech. J.*, **40**(1), 65–84.

Mortensen, R. E. (1975). Book review of "System Theory—A Unified State-space Approach to Continuous and Discrete Time Systems," *SIAM Rev.*, **17**(4), 699–703.

Oppenheim, A. and Willsky, A. (1983). *Signals and Systems*. Prentice-Hall, Englewood Cliffs, N.J.

Orfandis, S. J. (1985). *Optimum Signal Processing: An Introduction*. Macmillan, New York.

Papoulis, A. (1965). *Probability, Random Variables, and Stochastic Processes*. McGraw-Hill, New York. (Note: A revised edition was published in 1984.)

Rugh, W. J. (1981). *Nonlinear System Theory*. Johns Hopkins University Press, Baltimore, MD.

Vakman, D. E. (1968). *Sophisticated Signals and the Uncertainty Principle in Radar*. Springer-Verlag, New York.

Wiberg, D. M. (1971). *State Space and Linear Systems*. Schaum's Outline Series, McGraw-Hill, New York.

Williams, C. S. (1986). *Designing Digital Filters*. Prentice-Hall, Englewood Cliffs, N.J.

Wong, E. and Hajek, B. (1985). *Stochastic Processes in Engineering Systems*. Springer-Verlag, New York.

Index

Admissible random variable, 4
Algebra of events, 3
Ambiguity function, 164, 169
Analytic signal, 161
Autocorrelation, 68, 84, 94, 97, 98
Autocovariance, 63, 82, 84, 94, 103

Band-limited process, 92
Bandpass process, 147, 148
Bayesian assumption, 41, 192
Bochner's theorem, 71, 79, 86
Brownian motion, 124

Cauchy principal value, 161, 171
Causality, 58, 60, 77, 87, 190, 197
Channel, 111, 131ff, 156
Characteristic function, 6, 8, 29
Chi-squared distribution, 153
Completing the square, 26
Complex envelope, 166
Complex r.v., 127
Conditional covariance, 40
Conditional density, 9, 34
Conditional expectation, 43
Conditional mean, 40, 43
Conditional probability, 9
Confidence level, 224
Continuous parameter process, 13
Continuous r.v., 5
Controllability, 180
Convolution, 63, 66, 149, 161ff
Covariance matrix, 24, 51, 61, 82
Cross covariance, 66, 84, 103
Cross spectral density, 74, 75, 85ff
Cumulative distribution function, joint, 8

Delta function, 69, 90, 100, 136
Density function:
  Gaussian, 21, 24, 31
  probability, 5
Difference equation, 71, 78, 173ff
Differential equation, 104ff
Discrete parameter process, 13, 63
Discrete r.v., 5
Disjoint sets, 3
Dispersion, 166
Distribution function, 4

Eigenfunction, 120, 140
Eigenvalue, 117, 140
Eigenvector, 117, 120
Empty set, 3
Energy spectrum, 98
Ensemble, 52, 54, 81
Ergodicity, 93ff
Error covariance, 43, 202, 205
Estimate:
  Bayes, 41
  maximum likelihood, 218
Estimation theory, 41, 218

Event, 2, 4, 55
Expectation operator, 52, 63, 93
Expected value, 6, 8

Filter:
  bandpass, 132
  low-pass, 146, 148, 153
  matched, 99
  noncausal, 99
Fourier series, 69
Fourier transform, 7, 74, 85, 87, 98, 137
Function of r.v., 8

Gaussian process, 15, 49ff, 62, 81ff
Gaussian r.v., 21ff, 31ff
Gram-Schmidt orthogonalization, 13, 197

Hilbert space, 10, 115, 141
Hilbert transform, 161

i.i.d. sequence, 51
Impulse response, 87, 100, 103, 132, 153
Independence, 9, 11, 53
Independent r.v., 9, 24, 50
Inner product, 11, 116
Innovations process, 193, 200
Integral equation, 117, 140
Intersection, 9
Inversion formula:
  Fourier integral, 7, 69, 85
  matrix, 34, 38

Jacobian, 26, 176
Joint characteristic function, 8, 29
Joint cumulative distribution, 8
Joint density, 8

Kalman filter, 189ff
Karhunen-Loeve expansion, 122

Landau and Pollack, 140
Laplace transform, 87
Linear operator, 72
Linear system, 71, 87
Low-pass signal, 147

Marginal density, 9
Marginal integration, 9, 33
Markov process, 14, 177
Matched filter, 99, 101
Matrix exponential, 106
Mean, 6, 21, 50, 61, 63, 82, 94, 103
Memoryless nonlinearity, 156
Mercer's theorem, 120
MMSE criterion, 41, 192
Modulated signal, 93
Moments, calculation of, 6
Monte Carlo simulation, 17, 52, 93, 217ff

Narrow-band approximation, 148
Narrow-band noise, 92
Nonlinear system, 145ff
Non-negative definite function, 86
Norm, 11, 116
Normal distribution, *see* Gaussian r.v.

One-step prediction, 198, 202
Operator, 62, 72, 115, 117
  compact, 118
  self-adjoint, 118
Orthogonal decomposition, 190, 197
Orthogonality principle, 206
Orthogonalization, Gram-Schmidt, 13, 197
Orthogonal random variables, 13, 122
Orthonormal basis, 118

Paley-Wiener criterion, 91, 101, 148
Periodic process, 128
Positive definite matrix, 31, 51, 213
Power set, 3
Power spectral density:
  for continuous time process, 85ff
  for discrete time process, 68, 75
Probability:
  measure, 3, 58
  trio, 3, 58
Prolate spheroidal functions, 228
  propagate formula, 198ff

Radar:
  ambiguity function, 164, 169
  uncertainty principle, 165
Random:
  experiment, 2
  process, 13, 81
  sequence, 53, 173ff
  variable, 4
  vector, 4
Recursive filtering, 196, 198

Sample:
mean, 53, 219
space, 2, 58
variance, 53, 221
Sampled-data transfer function, 74
Saturation nonlinearity, 157
Schwarz inequality, 12, 20, 168
Second-order process, 83
Second-order random variable, 11
Signal to noise ratio, 135
Simulation, computer, 17, 52, 93, 217ff
Spectrum, 119
Standard deviation, 222
Standard Gaussian density function, 22
State space, 14, 91, 104, 182
State vector, 105
Stationary process, 84, 93
Stationary random sequence, 62
Statistic, 2, 41, 95, 218
Step function, 132, 136, 159, 160
Stochastic process, 13, 81
Stochastic realization problem, 77, 92
System weighting pattern, 112

Taylor series, 155
Time average, 94, 97
Toeplitz form, 176
Transfer function, 74, 87, 99, 132
Transition matrix, 107
Triangle inequality, 159, 168
Triangular factorization, 32, 51, 60, 209ff

Unbiased estimate, 95, 219
Uncertainty principle, 136ff, 148
Uncorrelated random variables, 24
Union, 18
Update formula, 196

Variance, 21, 50
Vector, 8, 31, 55, 115
Vector-valued white noise, 108, 111
Volterra series, 155

Waveform, 81, 93, 97, 139
White noise, 51, 90, 108
Wiener process, 124
Wiener-Volterra series, 155

Z-transform, 71